Integrative Cardiovascular Chinese Medicine

A Prevention and Personalized Medicine Perspective

Integrative Cardiovascular Chinese Medicine

A Prevention and Personalized Medicine Perspective

Anika Niambi Al-Shura, BSc., MSOM, Ph.D

Niambi Wellness
Integrative Cardiovascular Chinese Medicine
Tampa, FL

ELSEVIER

AMSTERDAM • BOSTON • HEIDELBERG • LONDON
NEW YORK • OXFORD • PARIS • SAN DIEGO
SAN FRANCISCO • SINGAPORE • SYDNEY • TOKYO
Academic Press is an imprint of Elsevier

Academic Press is an imprint of Elsevier
The Boulevard, Langford Lane, Oxford, OX5 1GB
525 B Street, Suite 1800, San Diego, CA 92101-4495, USA

British Library Cataloguing-in-Publication Data
A catalogue record for this book is available from the British Library

Library of Congress Cataloging-in-Publication Data
A catalog record for this book is available from the Library of Congress

ISBN: 978-0-12-420014-2

For information on all Academic Press publications
visit our website at store.elsevier.com

Typeset by MPS Limited, Chennai, India

14 15 16 17 18 10 9 8 7 6 5 4 3 2 1

In Memory

This textbook is dedicated in memory of my late parents. To Mary A. Cummings (1954–2006) who passed away of heart failure, I kept my promise of finding out what was wrong with you. To my father Abdur- Rahman Qurban Al-Shura (1949–1980) I followed what you advised me to do in life, and remembered that you taught me to always find a way to make it happen.

Dedication

The energy and effort behind the research and writing of this textbook is dedicated to my son, Khaleel Shakeer Ryland. May this inspire and guide you through your journey in your medical studies, career and life.

Special Thanks

Dr. Claudio Violato
Director of International Graduate Medical Education
University of Calgary School of Medicine and Universite Ambrosiana

I thank you for your guidance as my doctoral supervisor on the thesis of developing Personalized Integrative Cardiovascular Chinese Medicine.

Dr. Chen Xiao Hu, MD, PhD
Deputy Dean of Department of Cardiology
Professor of Integration of Chinese medicine and Western medicine
Nanjing University of Chinese Medicine

Chairman of Jiangsu Provincial Professional Committee of Microcirculation

Vice-Chairman of the Cariovascular Committee of Jiangsu Association of the integration of Chinese Medicine and Western medicine

Vice-Chairman of Hypertensive Cardiovascular Disease Committee of Chinese Medical Doctor Association

I appreciate your insight on cardiovascular diseases during proof editing.

Professor Wang Shou Chan
Director of the Institute of Pediatrics
Nanjing University of Chinese Medicine

President of the Pediatric Specialty Committee of the World Federation of Chinese Medicine Society (WFCMS)

I thank you for your influence in appointing me as a Vice-Chairman of the Pediatrics Sub-Committee (WFCMS).

Dr. Carol Cummings, PhD
Professor of Psychology and Pan-African studies
University of Louisville
Louisville, KY

I appreciate your insight and experiences on human behaviors which helped to shape the direction of the Practitioner Empathy and Patients Perspective Chapters.

Dr. Zhu Xian Hui
Physician and Doctoral Candidate
Department of Cardiology
Nanjing University of Chinese Medicine

I appreciate your assistance and advice on the Hypertension chapter.

Special thanks to **Dr. Sunil K. Trevedi**
Chief Research Officer at the U.N. Mehta Institute of Cardiology & Research Centre.

I appreciate you for your suggestions on the direction of this textbook.

Illustrators

Samar Sobhy and Shaza Ashraf Yassen Shabana

Contents

Part V
Vascular Diseases

Acknowledgements

A special acknowledgement to my 7 year medical students at Tianjin Medical University (2012-2013) who served as cardiovascular research assistants. I thank you for the hours of research into cardiovascular diseases, diagnosis and treatment using the integration of Chinese medicine and Western medicine. I appreciate you sorting through thousands of pages of Chinese medical literature, the translation into English and copy editing. May your future medical careers be successful.

An Qi He
Bin Lin Da
Han Jiang
Chen Hua
Jia Ying Luo
Jun Zhang
Lin Lin
Ming Lu
Nang Zhang
Ping Tang
Hu Si Le
Zhao Tian Man
Wen Xing Ning
Xing Wen Zhao
Tang Ying Mei
Li Ying Ying
Xiong Yong Qin
Ding Yu
Li Yan Jun

PART I

Personalizing Medicine

Chapter 1

Holism

Chapter Objectives

After studying this chapter you should be able to:

1. Describe the aspects of yin and yang, essence, qi and blood as they contribute to normal health
2. Explain the five-element cycle as evidence of yin and yang characteristics in nature, its correspondence to body organs and the normal and pathologic cycles
3. Describe the anatomy and pathophysiology of the heart and the interrelationship with other body organs according to Chinese medicine and Western medicine
4. Outline the base model for holistic medicine as it relates to the roles of the health practitioner and patient
5. Determine practitioner characteristics in humanistic medicine
6. Summarize the context for holism in personalized medicine

OVERVIEW

This chapter briefly explains the holistic basis for the development of integrative cardiovascular Chinese medicine. The purpose is to bridge specialized patient responsibility, with a specialized means for personalizing medicine and health care. The basic theories, medical sciences and scientific principles of traditional Chinese medicine (TCM) and Western medicine blend to construct the three main aspects within this system.

The first aspect includes the perspectives of the human body. In both TCM and Western medicine, the organs/viscera and organ systems are interconnected and interdependent. In TCM the opposing aspects of yin and yang are used to describe the intricate and delicate balance of physiologic properties and biochemical products which function to maintain homeostasis. In addition, TCM describes the specific and inevitable characteristics of earthly elemental resources and the forces of nature during the four seasons. The human body

Integrative Cardiovascular Chinese Medicine. DOI: http://dx.doi.org/10.1016/B978-0-12-420014-2.00001-6

must adapt to these environmental factors through dietary and lifestyle choices, or face pathologic consequences which match those factors.

The second aspect focuses on the interconnection between the cardiovascular system and other organs/viscera. In TCM, the heart functions in a cycle with the spleen, lungs, kidneys and liver – which are collectively called the zang organs/viscera. Using the imagination, their vital and pathologic functioning symbolically relates, respectively, to the characteristics of the earthly elemental resources of fire, earth, metal, water and wood. The story of the relationship dynamics between the zang organs is then explained in detail through the normal and pathologic cycles. The theories of yin and yang, essence, qi and blood explain their importance as 'vital substances' in the innate development, physiologic functioning and aging milestones of human life. The vital substances are also factors that drive the cycles among the zang interrelationships. The zang organs and their interrelationships are described in the theories of the five elements. In Western medicine, the cardiovascular and other organ systems provide most of the biochemical and processing requirements for the entire body. As the zang viscera relationship is described in TCM, in Western medicine the relationship between the viscera is expressed through the results of movement of blood, nutrients, biochemicals and other substances through the vessels. The results are tangible symptoms and signs experienced under the influences of illness, recovery and vitality, and palpable on physical examination.

The third major aspect includes the practical aspects of holism. The patient has a direct responsibility towards his or her own well being. In this context, competency can be seen in their mental outlook, lifestyle choices and outward behavior. An actively responsible patient can thrive on prevention practices with guidance from health practitioners, and, when necessary, they can be recruited to cooperate in a treatment plan for recovery. The holistic practitioner's primary role should be as humanistic guide and consultant, with a minor role in intervention and recovery. The knowledge, skills and behaviors are often influenced by the scientific principle and reductionist practices, no matter what the system of medicine. Harmonious collaboration between patient and practitioner can assist in the development of personalized medicine.

Key Points

- The five zang viscera are yin types because they store qi, blood and or essence. These include the heart, lungs, spleen, kidney and liver.
- The theory of the five elements in Chinese medicine can be considered as medical evidence of yin and yang theory. The natural elements correspond symbolically to matters in health state, promotion and disease manifestation.
- The brain is connected to the heart, and the marrow promotes clear thinking, speaking, emotional and psychologic stability. It governs the qi action for heart animation, motivates all zang organs in the production of qi, blood, body fluids, biochemicals and other substances for life functioning.

It works together with innate essence to nourish the zang and fu viscera and other tissues.

- Qi is the basic energy for life activity, which animates existence and promotes the physiologic function and metabolic processes. There are four types of qi: primordial, pectoral, nutritive and defensive. The brain absorbs these and generates qi for nerve signaling and regulation of senses, excretion and the conduction system.
- The heart shares a relationship with the spleen, liver, and lungs in the production and circulation of blood. In the relationship between heart and spleen, the spleen generates and controls the blood and the heart governs its circulation. The heart and liver share a relationship concerning blood. The heart circulates the blood and the liver stores and regulates the flow. The heart and lung relationship concerns the movement and circulation of blood.
- The heart qi motivates the circulation of blood, and lung qi purifies it with fresh air; respiration also assists with circulation.
- The heart and kidneys share a relationship of the dependence of essence and water (yin) on spirit and fire (yang).
- The heart intrinsically pumps and circulates blood and other substances throughout the closed circulatory system at the average rate of 70–80 beats/minute. The conduction system of the heart is the source of heart qi which regulates the movement of blood, the dilation and contraction of the vessels and the transmission of the pulses.
- In Chinese medicine, the heart has the physiologic functions of governing blood and dominating blood vessels, and the psychologic function of housing the inner spirit, which is associated with mental activity and the emotion of joy.
- The heart and liver share a relationship concerning blood and the mind. The heart governs and dominates brain activities of emotions, intelligence and psychology. The liver governs the free flow of qi which is necessary for the smoothness of emotions, blood and the mind. If the heart blood is deficient and the liver fails to smooth the free flow of qi, then emotional and psychologic issues will appear with symptoms such as depression, irritability and hair-trigger anger.
- A base model for integrative cardiovascular Chinese medicine is to blend the scientific method and reductionist perspectives to cultivate objective holistic medical practices that collaborate with a healthy patient mental and lifestyle choice.

PERSPECTIVES OF THE HUMAN BODY

Yin and Yang Theory

In TCM, yin and yang theory describes the basic opposing aspects within the structures and functioning of the outer universe (heaven) and surrounding nature (earth). The yin and yang symbol is a circle with one half dark, representing yin; the white half represents yang. Within the dark yin half is a small circle of white and within the yang white half is a small circle of black. They represent the potential for balance through interdependence, mutual restriction, reciprocity and mutual transformation. Examples of yin attributes include earth, entering, descending (lower), darkness, coldness, dampness, femininity, yielding, static, internal, material formation. Examples of yang attributes include heaven, exiting, ascending (upper), illumination, warmth, dryness, masculinity, reaping, moving, external and immaterial.

Yin and Yang in Nature

The daily attributes of yin and yang are experienced in a 24-hour cycle and during the four seasons of the year. In a 24-hour cycle, the day begins with the interdependence of yin and yang as the moon descends and the sun ascends. With the transit of the sun across the first half of the sky in the morning and early afternoon, the day becomes brighter and yang is dominant. With the transit of the sun across the second half of the sky in the late afternoon and early evening, the day becomes dimmer as yang brightness transforms with a touch of yin darkness. As the sun sets and yang descends, yin ascends. With the transit of the moon across the first half of the sky in the evening and night the day becomes darker and yin is dominant. The phases of the moon can be characterized as yang for the waxing cycles and yin for the waning cycles. With the transit of the moon across the second half of the sky in the late night and early morning the sky becomes dimmer, as yin darkness transforms with a touch of yang brightness. As the sun rises, a new day begins with a mixture of yin darkness and yang light. It is interesting that pregnant mothers have Braxton–Hicks contractions or go into labor during full moon cycles. Also, minor and serious illnesses, myocardial episodes and death seem to occur between the late night and early morning.

During the 12 months of the year, especially in geographic locations where there are four seasons, examples of mutual yin and yang restriction and reciprocity can be easily noticed with the transitions in nature. The first quarter of the year begins with the yin coldness of winter. As spring emerges, the yang energy restricts the coldness to transform into warmth, which is optimal for gestation of life. During the second quarter of the year, yang energy gradually dominates as spring warmth transforms into the heat and flourishing of life in the summer. During the third quarter of the year, yang energy is restricted by yin which cools summer heat and prepares for the reaping actions of the fall. In the fourth quarter of the year yin gradually dominates into dormant winter. It is interesting that heart problems are exacerbated during winter and summer months.

Yin and Yang and the Human Body

In medical science the human body is divided into anterior and posterior, proximal and distal, superior and inferior, internal and external, internal organs, tissues and their physiologic functioning. All can be correlated with yin and yang attributes. For example, in Chinese medicine, all the organs of the trunk are divided into yin and yang types and are called zang and fu viscera (organs). The five zang viscera are yin types because they store qi, blood and/or essence. These include the heart, lungs, spleen, kidney and liver. The six fu viscera are yang types because they transport and transform food nutrients and waste products for distribution and excretion. These include the small intestine, large intestine, stomach, urinary bladder and gall bladder and the san jiao. Other organs, such as the brain, lymphatic, endocrine and reproductive organs, have a special connection in this interrelationship. Each zang and fu viscus has yin and yang aspects as related to their position in the body, their anatomic structure, physiologic function and interrelationship with other body organs. They also have a holistic matching relationship. The heart zang is matched to small intestine fu, the spleen zang is matched to stomach fu, the lung zang is matched to large intestine fu, kidney zang is matched to bladder fu, and liver zang is matched to gall bladder fu. Within each viscus, physiologic functioning is primarily yang, with subfunctioning within organs and organ systems which can be both yin and yang according to relationship. Biochemical products – such as hormones, enzymes, metabolites, pigments, vitamins and minerals, sugars, ketones, polysaccharides, proteins, fats, body fluids, blood, etc. – are primarily yin. They also have subproperties which can be classified as yin or yang according to their pH, concentration, mechanism of action and activation of vital physiologic functioning of organs and tissues.

PATHOLOGY: VITAL QI VERSUS PATHOGENIC QI

Vital qi is an energetic health state of the body, which is associated with yang. Important tissues, structures, fluids, biochemical products and all of their functions move within their unique yin and yang cycles to maintain homeostasis. As the body is constantly evaluating the ability to resist or endure and recover from an attack of disease, maintaining the balanced level of the yang energetic state is crucial (see Fig. 1.1). It depends on sufficient levels of yin substances, such as blood, body fluids, vitamins, minerals, other nutrients and biochemicals, which nourish and protect proper levels of yang so that neither is depleted. Pathogenic qi is considered an invasive factor in the health state where lowered immune resistance leads to health deterioration. It is also divided into six factors according to yin and yang: yin pathogenic factors are coldness and dampness, while yang pathogenic factors are wind, dryness, heat and fire. It moves from the outside to the inside, penetrating the first line of defense to move towards the second line. Changes in weather, seasonal allergies, epidemics/pandemics, sexually transmitted diseases, etc. are examples of pathogenic qi. In a very basic sense, the severity and magnitude of the pattern in which pathogenic qi attacks and affects the body is from extreme yin (cold)

FIGURE 1.1

FIGURE 1.2

to extreme yang (fire). Over time, untreated or under-treated exogenous pathogens can cause organ damage and organ failure.

When yin is deficient and unable to restrict the action of yang, the result is a yang excess (as shown in Fig. 1.2). The pathogenic factors are characterized by a diminishing of moisture and nourishment of the tissues and organs, resulting in fluid deficiency and dryness. The therapeutic principle is to restore balance by nourishing yin and stabilizing or subduing yang.

When yang is deficient and unable to restrict the action of yin, the result is a yin excess (as shown in Fig. 1.3). The pathogenic factors are characterized by fluid retention and inability of the body to maintain warmth. The therapeutic principle is to restore balance by stabilizing or reducing yin while warming and tonifying yang.

When the vital qi is damaged, both yin and yang are deficient (as shown in Fig. 1.4) due to the excess demands of mutual restriction. Yin is damaged because of the inability of yang to warm and create a steaming effect to

FIGURE 1.3

FIGURE 1.4

cause a kind of condensation which assists in producing fluids. Yang is damaged because of yin's inability to reciprocate by sharing the fluids to nourish and regulate yang levels. The therapeutic principle is to nourish yin while tonifying yang.

FIVE-ELEMENT THEORY

The theory of the five elements in Chinese medicine describes the activities of yin and yang as they are found in nature (Table 1.1). Each element is characterized by its personal attributes, which are coincidentally noticed in the health and pathologic states in humans. The patient and holistic health care practitioner can adapt this system to personalize the diet, make lifestyle adjustments and create a treatment plan.

TABLE 1.1

Element	Element Characteristic
Fire	Warmth, heat, flaring upwards
Earth	Yielding, receiving, creating, generating, sowing and reaping
Metal	Descending, astringing, clearing, changing
Water	Moistening, coldness, flowing downward,
Wood	Flexibility, strength, flourishing, developing upward

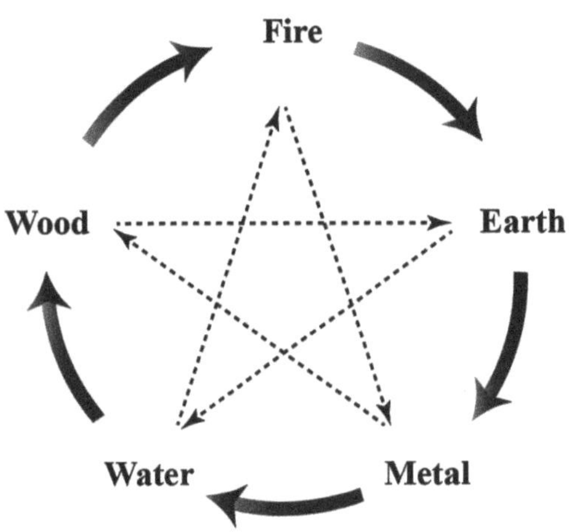

FIGURE 1.5

Normal Cycle

The normal cycle represents balance and harmony among the elements (Fig. 1.5). One element helps with the development of another through promoting or restricting. Two aspects within this are the mother and child promoting cycle and the grandmother and grandchild restricting cycle. In the promoting cycle, the fire mother warms and transforms the earth child, the earth mother generates and nourishes the astringent property of metal child, the metal mother nourishes the nutrient fluid properties of the water child and the water mother nourishes the flourishing growth of the wood child. The restriction cycle is an authority and subordinate cycle which is also called the grandmother and grandchild cycle. One element restricts the actions of another as a means of reinforcing the promoting cycle. The wood grandmother restricts the over-expansion of the earth grandchild, the fire grandmother restricts the over-astringency of the metal grandchild, the earth

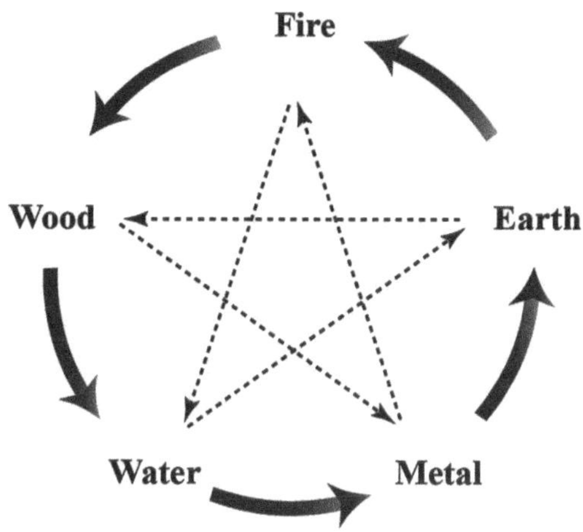

FIGURE 1.6

grandmother restricts the over-flow of the water grandchild, the metal grandmother restricts the overgrowth of the wood grandchild and water grandmother restricts the flaring blaze of the fire grandchild.

Pathologic Cycle

The counter action and violation cycles are pathologic states. An imbalance between two elements anywhere in either of the two cycles triggers a domino effect of failure within the entire five-element cycle. In the counter action cycle the child element becomes excessive, uncontrollable, and then turns on the possibly deficient mother element. For example, the blazing fire child burns the wood mother to ashes (see Fig. 1.6). The overgrowth of the wood child consumes all of the nourishing water mother, the water child floods and corrodes the metal mother, the excessive metal child poisons the earth mother, the over-generating earth child smothers the fire mother.

In the violation cycle, the insubordinate grandchild element becomes excessive and overpowers the grandmother element. The over-generating earth grandchild swallows the flourishing control of the wood grandmother. The blazing fire grandchild dries up the moistening control of the water grandmother. The over-flourishing wood grandchild depletes the nutritionally astringent control of the metal grandmother, the over-flowing water grandchild floods out the earth grandmother.

ESSENCE, QI, BLOOD AND BODY FLUIDS

Essence

Essence is described in two aspects which include the innate from genetic factors and the acquired from lifestyle choices of nutrition.

First, there is innate essence – which is inherited from the parents during the process of meiosis after conception. The level of thriving energy of the developing embryo becomes the inner drive and spirit which mixes with, and strengthens, the innate essence to motivate survival in the fetus during the gestational period to birth. The lifetime supply of innate essence is then stored in the kidneys after birth to assist with further development. Between adolescence and adulthood, the essence develops into tankui which brings puberty and prepares males and females for adulthood and as potential genetic donors to offspring. Essence also nourishes the brain and the bones. It bathes in the fluid around the brain and spinal cord to promote clear thinking, speaking, emotional and psychologic stability. In the bones and teeth, it helps to form marrow, and provides sufficient nutrients, strength, flexibility and durability. In addition, marrow in the long bones is the source from which blood is produced.

Secondly, acquired essence is made up mostly of water and gastric enzymes mixed with vitamins, minerals, carbohydrates, amino acids and other nutrients derived from the consumption and digestion of food. It travels in the bloodstream as nourishment and balances the maintenance of other biochemical mechanisms. Also, it assists the innate essence around the brain and spinal cord. Acquired essence is crucial for supporting life.

Deficiency of congenital essence results in cognitive and developmental disabilities in the fetus and infant. Deficiency of kidney essence during adolescent development results in poor intellectual ability, forgetfulness, dizziness, and a weak and unhealthy body constitution. During the phase middle age to senior adulthood, essence deficiency results in kidney deficiency and diminished brain functioning in the stages of senility. Acquired essence primarily becomes the substance which is transformed into the qi of the body – which provides energy for tissues, organs and physiologic functioning.

Qi

Qi represents the basic energy which animates regular life processes and is created partially from acquired essence. Qi warms, like yang energy, and spreads blood to moisten the skin surface to allow anti-pathogenic qi to remain on active duty. Qi also assists in maintaining the peripheral body temperature, proper physiologic functioning, internal blood circulation and secretion of body fluids. When this qi is abundant, the body can maintain normal body temperature levels. Temperature regulation is also involved in the normal body transformations. Food is divided into clear and turbid. The clear food, as nutritious substances, transforms into catalysts for metabolic reactions. Turbid substances include indigestible food components and waste synthesis of biochemical products. There are four types of qi: primordial, pectoral, nutritive and defensive.

Qi moves in four directions. Two of the directions are related to the motion of creating the acquired qi. Entering motion moves qi from the exterior to the interior – such as fresh air into the lungs to become lung qi, and processed food from the stomach to duodenal and small intestinal absorption to become nutritive qi. Once acquired essence from food mixes with acquired qi from breathing in air, the transformation into qi occurs in the vessels, and the blood carries all necessary nutrients and substances to the organ systems in the body. The other two directions are related to the normal movement of the qi once it animates zang or fu viscera. Ascending motion moves qi from the lower parts of the body to the upper parts, such as liver qi and spleen qi. Descending motion moves qi from the upper parts to the lower parts such as in heart qi, lung qi, stomach qi, small intestine qi, large intestine qi and bladder qi. According to the zang or fu viscera or organ system, qi provides the animation to perform six functions: to defend, control, propel, warm, nourish and transform.

(a) *Primordial qi.* This qi is really the action of innate essence as it activates development between puberty and adulthood.

(b) *Pectoral qi.* This qi is acquired by breathing fresh air into the lungs, and nutritive qi is acquired from food. It accumulates in the thorax region, the sea of qi area, and moves through the lungs and heart through blood circulation. It functions to promote and support the actions of the primordial qi, to move the lobes of the lungs, and to circulate the blood.

(c) *Nutritive qi.* This qi is the nutrition formed from food essence. It assists in the generation and production of blood, then circulates the blood to nourish the tissues, organs and organ systems.

(d) *Defensive qi.* This qi is a yang action, also called anti-pathogenic qi. It is created from pectoral qi and nutritive qi. Defensive qi protects the body, controls the temperature and the secretion of sweat. It spreads around the body by first circulating over the body surface to warm the skin and hair to make them lustrous. Next, it moves to the interior to contribute to the immune system and organ system activities. Anti-pathogenic qi is the motivation for the immune system, especially for the first and second line of defense. Various lymphocytes and leukocytes react with and kill microbes.

When this qi is healthy and abundant, the body is able to defend itself and prevent illnesses, or to recover quickly should an illness occur. When this qi is deficient, the immune responses become delayed and eventually diminished, causing exposure to infection.

(e) *Controlling qi.* Controlling qi functions to regulate the discharge and retention of fluids. It allows opening and closure within the sphincters of hollow organs to adjust the rate of secretion of urine and feces. It controls dilation and contraction of the cervix for menstruation and childbirth, skin pores for secretion of sweat, valves in the blood vessels for blood flow, and glands for saliva and semen. Deficiency of this qi will lead to profuse sweating, hemorrhage, hypersalivation, spermatorrhea and nocturnal emission, leukorrhea, incontinence, diarrhea due to rectal prolapse, irregular vaginal bleeding or menstruation and miscarriage.

(f) *Propelling qi.* Propelling qi is the energy which is involved in the physiologic functions of the meridians, tissues, organs and organ systems, as well as biochemical mechanisms and products. For example, the autonomic nervous responses are responsible for the beating of the heart, the movement of the lungs and the eyelids, peristalsis for movement of feces and secretion of urine. When this qi is abundant, movement of qi through meridians, physiologic function of organ systems, body fluids and other biochemicals are normal. Deficiency of this qi contributes to pathology, and to developmental disabilities in the fetus and the adolescent.

Blood

In Chinese medicine, blood is a yin substance. The basic components are (i) essence from the kidneys, which enters the liver to nourish the stored blood before it is dispersed, and (ii) nutrition from the transportating and transformating function of the spleen and stomach. At proper levels, and with the correct composition, the blood nourishes the body, making the complexion clear, the hair and nails lustrous, the muscles and tissues strong, sinews and joints flexible, and the intellectual and psychologic abilities clear, rational and powerful; it also enables the reproductive systems and cycles to function normally. Blood is moved by the propelling actions of qi; this causes the movement of nutrition and waste products throughout the vessels and within tissues and organ systems. Proper circulation is contingent upon the smoothness of the vessels, the temperature and the quality of blood. The spleen, liver, lungs and heart are important viscera in the production and circulation of blood.

The spleen. Spleen qi controls the formation and the steady movement of blood through the vessels.

The liver. The liver stores and controls the volume and the free movement of blood.

The lungs. The lung qi mixes fresh air with returning blood to form pectoral qi which is then moved through the heart.

The heart. The heart qi propels the blood through the vessels out to the rest of the body.

Body Fluids

Body fluids are considered to be the interstitial fluids, saliva, tears, and gastric juices. They moisten the tissues, muscles, body organs and skin. In Chinese medicine, the production of these fluids is influenced by proper gastrointestinal function by the spleen and stomach qi. These vital fluids are formed from the food we eat and the water we drink. The lungs control respiration, which releases moisture and distributes body fluids through full breathing. Body fluids are excreted through the pores as sweat. They are also dispersed throughout the body by the action of lung qi. The qi energy motivates the circulation of lymph and interstitial fluids, which is sent to the kidneys for filtration, absorption and reabsorption. Fluids are then transformed and sent to the bladder as urine, or collected in the large intestine to assist the discharge of feces.

THE INTERCONNECTIONS OF THE CARDIOVASCULAR SYSTEM

The Human Body

In Chinese medicine, the human body is considered to be a whole and functioning unit of separate but interdependent parts. Some of this is described in the five-element visceral theories of the zang and fu organs. The five zang organs include the heart, spleen, lungs, kidneys and liver; these organs are considered to be yin viscera because they are the main venues for life sustenance. They have a respective relationship with the five fu organs, which include the small intestine, the stomach, the large intestine, the bladder and the gall bladder (as shown in Fig. 1.7). These are considered yang viscera because they are hollow vessels for transforming nutrition and for transporting waste material. This respective visceral relationship functions in a compensatory cycle for maintaining balance within and throughout the body. Therefore, the balance of the relationship is interconnected with other organ systems and it influences the physiologic mechanisms of tissues, vessels, orifices, senses, blood and other body fluids, essence, and qi energy. Imbalance within this relationship contributes to pathology.

In Western medicine, the human body is also considered as a whole and functioning unit of separate, yet interdependent parts. In anatomy and

FIGURE 1.7

physiology the body is divided into 11 balanced organ systems: the cardiovascular system, lymphatic system, nervous system, endocrine system, respiratory system, digestive system, urinary system, reproductive system, skeletal system, muscular system and integumentary system (see Fig. 1.8). The main organs include the brain, heart, lungs, kidneys, gastrointestinal tract (involving the stomach), the large and small intestines, and the hepatobiliary system involving the pancreas, liver and gall bladder. Other tissues, vessels, substances such as blood, hormones, body fluids, etc., and electrical

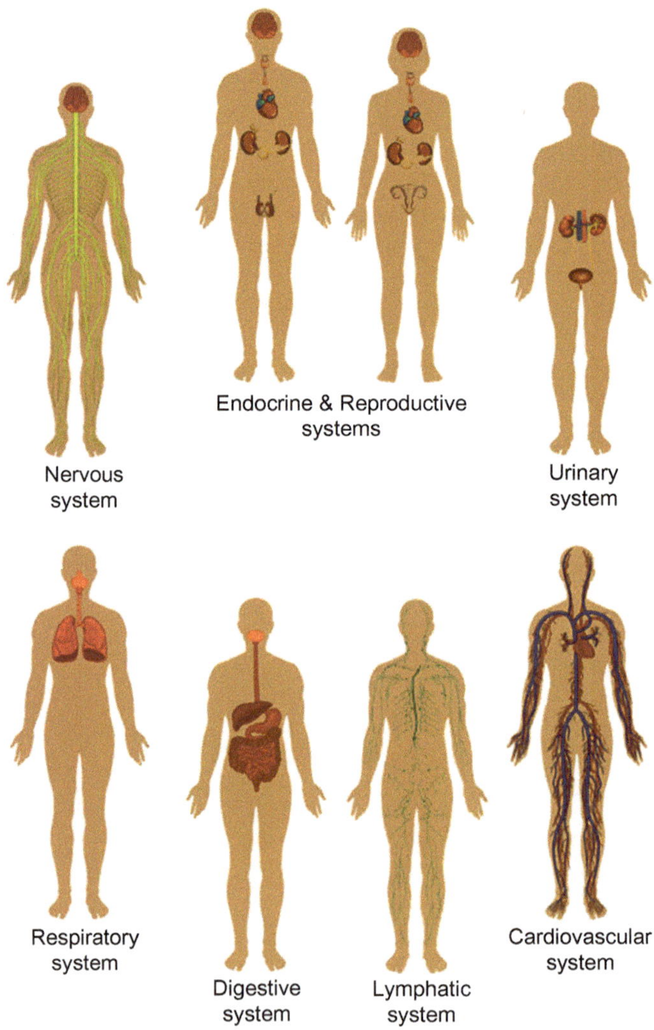

FIGURE 1.8

energy are produced, controlled, transformed and transported within this balanced system. Imbalance within this relationship contributes to pathology.

The Heart

The heart zang organ is considered the monarch of all organs in the Chinese five-element system. In humans, the heart is located in the center of the thoracic region, posterior to the sternum. The pericardium is a fluid-filled sac that protects the heart. A person's heart is considered to be the size of their fist, has a mass of between 200 and 350 g, and consists of three layers of tissue. The outer layer is called the *epicardium*; this is in contact with the tissues of the pericardium. The middle layer is called the *myocardium*; this is composed of striated cardiac muscle tissue. The inner layer is called the *endocardium*; this comprises the chambers, valves and endothelial tissues of the cardiac vessels.

In Chinese medicine, the heart has the physiologic functions of governing blood and dominating blood vessels. The heart, other organs and organ systems are nourished by the yin fluids and yang products found within the circulating blood. The heart has the psychologic function of housing the *inner spirit*, which is associated with mental activity and the emotion of joy. Inner spirit is a product of the complex yin and yang balance of activities within the brain. Physiologic brain activity is associated more with yin because of the presence and expression of biochemicals. Cognitive functioning involves thought and emotion and is therefore immaterial and associated more with yang. Inner spirit controls heart qi through the balance of yang. It is noticed as clear and quick thinking, being grounded, responding rationally, and having clearness of consciousness. The dilation and constriction of the blood vessels are also dominated by heart qi. The heart state can present as a physical manifestation on the face, the tongue and through the secretion of sweat. The face will be clear, and the skin will be lustrous as a result of blood flow through the capillaries. The tongue will be red, flexible and moist, providing the ability to speak with clear speech and eloquence. The secretion of sweat will be produced in sufficient amounts according to physical activity and environmental temperature.

Heart Qi

In Chinese medicine, has a function of governing blood which involves the movement and circulation throughout the body, and the continuous movement is contingent upon the robustness of heart qi.

The heart intrinsically pumps and circulates blood and other substances throughout the closed circulatory system at an average rate of 70–80 beats/minute. The conduction system of the heart is the source of heart qi which regulates the movement of blood, the dilation and contraction of the vessels and the transmission of the pulses. Heart qi is found within the cells of the myocardium, and is involved in contractility. Yin and yang balance is expressed through the phases of the resting and action potentials. The central nervous system controls baroreception, where sympathetic innervation leads to the release of norepinephrine to act upon B1 receptors of the sinoatrial node (SAN). The SAN is the site of manifestation and transmission of heart qi. Imbalance can lead to tachycardia, which can be characterized as an excess of heart qi, and bradycardia, which can be characterized as a deficiency of heart qi. Heart qi is also involved in contractility, cardiac output, and stroke volume, as they affect the amount and pressure of the blood – especially concerning left ventricular activity.

Heart Blood

The heart qi assists the circulation of blood throughout the body. The respiratory, endocrine, hepatobiliary, gastrointestinal and renal systems use the various types of vessel as reservoirs to excrete, exchange, and deliver products such as vital nutrients, fluids, hormones, other biochemicals and their waste into systemic circulation. These products comprise the measured composition, volume and quality of blood. Yin substances are water and other fluids; yang substances include oxygen, carbon dioxide, hormones, nutrients, biochemicals, microbes and other substances. Deoxygenated blood and the products return to the heart through the inferior and superior vena cavae. This deoxygenated blood, with yin and yang balance of products, is capable of influencing the direct functioning of the heart. For example, products from the renin–angiotensin system of the lungs and kidneys regulate fluid volume and blood pressure.

After the returning systemic blood enters the right atrium, it is then pumped into the right ventricle and sent through the pulmonary artery to the lungs. During gas exchange, some of the products remain in the oxygenated blood. As the blood flows through the pulmonary veins and re-enters the left heart, there is preparation for compensation and adaptation of pressure changes, stroke volume and potential cardiac output. The left ventricle will be most affected by these changes. From the left ventricle the blood then exits the endocardium through two major arteries. The cardiac vessels receive oxygenated blood to nourish the heart, and the aorta moves oxygenated blood back into the systemic circulation to repeat the holistic cycle of the heart.

THE FIVE-ELEMENT ORGAN/VISCERA SYSTEM (TABLE 1.2)

The Heart Zang and Small Intestine Fu

The heart and small intestine share the zang and fu holistic interior and exterior relationship on the meridian system. The heart shaoyin connects with the small intestine taiyang. Heart yang descends to the small intestines to warm and assist with the excretion of chyme during its role of transportation and transformation of food into nutritional essence in digestion. A portion of

TABLE 1.2

Element	Zang viscera	Fu viscera	Element characteristic
Fire	Heart	Small intestines	Warmth, heat, fire, spirit, governing, dominates/monarch of the zang and fu viscera
Earth	Spleen	Stomach	Yielding, receiving, creating, generating, sowing and reaping, producing, stabilizing
Metal	Lungs	Large intestines	Astringing, clearing, changing, purification
Water	Kidneys	Bladder	Moistening, storage, cold, coolness, essence
Wood	Liver	Gall bladder	Flexibility, strength, flourishing, storage, free flow

nutritional essence is then sent back to the heart which uses heat to change the color to red and becomes part of the components of blood.

The Heart Zang and Spleen Zang

The heart and spleen share a holistic interrelationship concerning blood, as the spleen generates and controls it and the heart governs its circulation. The spleen qi participation in transportation and transformation of food into nutritional essence contributes to the generation of blood. The heart governs blood, heart qi motivates the circulation, and spleen qi stabilizes the blood to remain inside the vessels.

The Heart Zang and Lung Zang

The heart and lungs share a holistic interrelationship concerning the movement and circulation of blood. The heart qi motivates the circulation of blood, and the lung qi purifies it with fresh air; also, respiration assists with circulation. Pectoral qi of the lungs is produced because of the mixing of nutritive qi and fresh air.

The Heart Zang and Kidney Zang

The heart and kidneys share a holistic relationship with the interdependence of essence and water (yin) with spirit and fire (yang). The kidneys represent yin and the heart represents yang. The heart houses the spirit and controls essence, which is stored in the kidney. The spirit is the energetic expression of essence, and essence is the material expression of spirit. The location of the heart in the upper jiao or region of the trunk means that it has a yang position, while the location of the kidneys in the lower jiao or region of the trunk means that it has a yin position.

The Heart Zang and Liver Zang

The heart and liver share a holistic interrelationship concerning blood flow and the mind. The heart circulates the blood, and the liver stores and regulates the flow. If the heart governs the circulation of a sufficient amount of blood, then the liver can store it. If the liver stores and regulates a sufficient amount of blood, then the heart can circulate it. The heart governs and dominates brain activities of emotions, intelligence and psychology. The liver governs the free flow of qi which is necessary for the smoothness of emotions, blood and the mind.

See Chapter 6 for a detailed explanation of specific organ functioning and the interrelationships in TCM and Western medicine.

THE PRACTICAL ASPECTS OF HOLISM

The scientific method suggests that truth is what the neutral observer seeks. Searching for the probability of an event or circumstance and reporting this result as an objective (neutral) conclusion is considered the norm. During medical scientific studies, researchers can design a qualitative or quantitative experiment, gather a sample population (ordinary people or target patients) and divide them into epidemiologic or physiologic groups. Later, the population can be mathematically analyzed and further subdivided into other distinct groups according to other requirements. Finally, results are written up and published as evidence of certain conclusions. On one hand this objective variation of neutrality leads to confusion over which information is most meaningful towards further analysis or a conclusion. On the other hand, one could argue that there are many facets of the truth, and reinforces the conclusion that there are no absolutes in regular life.

The public often relies on these scientific studies to give evidence of the general efficacy of methods, treatments and medicines. Practitioners rely on them to make crucial decisions in general patient management. These assumptions suggest that the way in which the published results would later be used was carefully considered beforehand. Further, it can be assumed that a thoughtful and educated brainstorming was done to derive a study question of any relevance, to design an experiment of any worth, and to report results of any merit.

Whether or not this is the case, published medical scientific studies have been the accepted method of determining suitability of any method, treatment or medicine. They are used for crucial decision making, and have become a major part of how medicine serves the public. Reductionism is a byproduct of the scientific method. For example, specializations in Western medicine devote educational and practice efforts to the pathology of an organ system. Research studies have led to much new knowledge about how the human body acquires, adapts, recovers and prevents diseases. Some reductionist advocates argue that Western medicine in its present state may not appear holistic, but it is. The study of all organ systems and their biochemical and pathophysiologic mechanisms includes how diseases are caused by intersystem failures. Therefore, diagnostic methods, implementation of treatment, follow-up and management, though devoted to a specific organ system, still deal with an aspect of holism.

Practitioners of TCM tend to distinguish it from Western medicine by claiming the term ‘holistic medicine.’ Within this term, one views the whole person, whole herb, whole meridian system, the intent of the practitioner, the method of insertion of the needle, the method of the tui-na stroke, etc. Many TCM advocates argue that the objective scientific method is nearly useless for studying most areas of Chinese medicine, especially acupuncture, and that it cannot be used to effectively measure or determine efficacy of methods, instruments and medicines within the domain of the holistic. The reason is that the results of diagnosis, treatment principles and treatments are usually more subjective in TCM.

Despite this, the scientific method has still been the major weapon that advocates of Western medicine have been using against TCM theories, practices and research results.

Most likely, the myriad published studies on nearly every area of TCM have involved scientific methods – presenting results and deriving conclusions which advance public acceptance.

However, if the scientific method in medicine also concludes with non-absolutes (because the population includes human lives), then all areas of Chinese medicine should be effectively evaluated using the scientific method. For standard scientific methods to gain acceptance, many clinical TCM practitioners may need to re-think how they perceive holism. Holistic practitioners who endeavor to provide care by integrating forms of Chinese medicine and Western medicine in cardiology can utilize truths revealed by scientific methods which can offer clues for providing personalized health care.

HOLISTIC MODEL FOR INTEGRATIVE CARDIOVASCULAR CHINESE MEDICINE (FIGURE 1.9)

Chinese medicine and Western medicine both share these qualities, which can be integrated for personalized medicine. Upon analysis, both appear to employ aspects of holism and reductionism. Examples of shared holistic

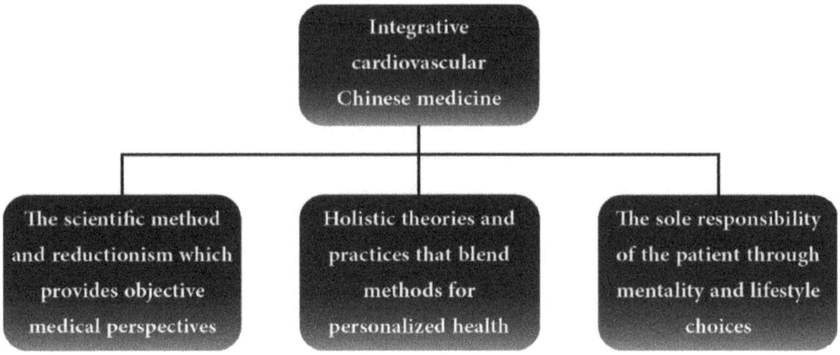

FIGURE 1.9

aspects include the basic medical sciences and theories, various physical examinations of the body (which include tongue and pulse analysis), diagnostic and treatment principles that consider several causes of a set of symptoms, treatment methods that include medicine, and lifestyle modifications. Chinese medicine detects organ system dynamics through advanced variations of five-element theory and differentiation of syndromes (DOS). Western medicine detects organ system dynamics and rules out other causes of symptoms through differential diagnosis.

These paradigms contribute to a prejudice shared by both types of practitioner: a requirement to declare a diagnosis, to wage a war to relentlessly pursue and destroy the direct cause, and to view with disdain every aspect of human existence that contradicts their ideals of perfect health. Consequently, the realization may be that research results and conclusions can determine the factors which cause disease in a set population, but may not be absolutely correct in predicting and determining what is true for all people. Personalizing medicine and health care means making decisions for the patient as a unique individual. It should include detailed factors the patient observes about themselves in their own daily lives, and in the clinic during the narrative and physical examination.

Ultimately, to make holistic health care more effective for the public, the practitioner must draw the line between the consequences of imposing their 'intent' and actually providing what the patient may need. Next, the practitioner would need to accept that patients often have set habits and lifestyle choices, so they must take sole responsibility for their own lives. The health practitioner would then need to establish and reinforce their primary role to educate, consult, and when necessary, provide intervention and recovery to as optimal a state as possible. From that point, if a patient is able, the responsibility should be relinquished back to them, so that the practitioner can resume their primary role. Consequently, the practitioner and the patient may need to share the same holistic outlook on life for recruitment and management to be successful.

Research programs and medical practices using integrative medicine may be able to use this model to guide a holistic clinical perspective.

HOLISTIC LIFE OUTLOOK

Responsibility for the Mentality: Thoughts and Emotions

In Chinese medicine, the heart governs the mentality: the mind, thoughts and emotions. These shape, and are influenced by, personal character. Personal character includes the innate genetic qualities mixed with learned values and mores from upbringing, culture, religion and life preferences. Striving for a good personal character is a sign of personal integrity. A healthy mentality can be observed in love, self-fulfillment and personal contentment. Challenges which test mental integrity happen during life choices and relationships with others. Thoughts and emotions are demonstrated through outward behavior. First establishing that successes and failures are part of regular life, and that neither must interfere with a balanced mentality, is a sign of a well-adjusted personal character. In addition, this outlook allows one to have a smoother recovery after adversity, which reinforces and continually builds personal character.

When the heart has a normal and strong ability to govern the mind and emotions, the person demonstrates honesty, gives and accepts truth, treats others with kindness and selflessness, honors obligations, admits faults, corrects personal mistakes, and abstains from greed and envy. In pathology, fear, worry, anger and guilt undermine personal character and the physiologic result is manifested in heart disease.

Responsibility for Lifestyle Choices

The lifestyle should include proper diet, physical activity, leisure activity, rest and sleep. For example, in Chinese medicine a proper lifestyle is vital for the normal functioning of the life forces of qi, blood, yin and yang. In both Chinese and Western medicine, a healthy diet should involve a variety of food choices and nutritional substances for proper sustenance. Physical activity means various exercises according to constitution and ability, but should also include hobbies, a fulfilling career and a healthy sexual life with mutual interest, love and affection. In addition, settings in nature during different seasons can provide a therapeutic venue for physical activities. Normal levels of rest and sleep are important, as this is the time when the body recovers from stress and illness.

When the vital forces of qi, blood, yin and yang are balanced, the body systems work in harmony and life is enjoyable. The heart qi and yang function properly by enduring fast activity, pumping a sufficient amount of warm yin blood, nutrition and hormones throughout the vessels. Also, the vessels will be open, clean, smooth and flexible to allow free flow of qi energy and blood, and the movement and removal of waste products. The blood and nutrients can nourish the organs to work properly to metabolize fat, sugars and other substances, and also bathe joints, sinews and muscles for free movement.

HUMANITY IN MEDICINE

It has been established that the primary role of the health practitioner should be, first, to do no harm. Next, to use their knowledge to promote and preserve good health, and finally to use their skills to intervene and return health to optimal levels. For the practitioner to be able to serve the patient in their basic role, the patient must be prepared to take personal responsibility for their own mental outlook and lifestyle choices, and the practitioner must have similar values in their medical morals. These morals should include sound education, skill and wisdom as a base for medical humanity. Practitioners should effectively communicate with the patient by listening and understanding the patient from their point of view. In personalized medicine, one method or treatment does not fit all. The clinical interview and patient narratives, physical examination and diagnosing can help to guide a patient's specific needs. Ultimately, the health practitioner must be good at connecting different factors to derive a diagnosis. Emotions, behaviors and lifestyle practices reflect stages of health and illness. During treatment and while predicting a prognosis, many practitioners will accept nothing less than perfection in the pursuit of optimal health. To facilitate this goal, it may be tempting to focus more on the disease, while ignoring the patient. Solutions may require expanding further understanding about pathology, lifestyle choices, body constitution, and physiologic characteristics of the constitution. Only by remaining steadfast in the practitioner's role and not overstepping the boundaries into a patient's personal responsibility can a health practitioner in any medical discipline or system truly begin to practice holism.

SUMMARY IN THE CONTEXT OF HOLISM FOR PERSONALIZING MEDICINE

For providing personalized care in integrative cardiovascular Chinese medicine, a mutual perspective in holism between the patient and practitioner is important. The participation and compliance of the patient is standard for treatment success. A patient must accept personal responsibility for their own lives, have control over their emotional outlook, adopt holistic lifestyle choices for prevention of disease, and actively participate in treatment and recovery periods. The health practitioner must accept remaining in their holistic role using their knowledge to guide and consult with limited and reasonable intervention and recovery duties, when necessary. For providing personalized medicine, more appropriate holistic knowledge as a guide, more efficient tools in intervention and a deeper wisdom and personal humanity are required. 'First do no harm' may mean not crossing boundaries and allowing interpersonal acceptance.

FURTHER READING

[1] Guirong L. Holism, systems theory and the study of the modernization of traditional Chinese Medicine. J Basic Med TCM 1996;2(3).

[2] Zhang J, Lu C. Philosophical origins of traditional Chinese medicine's holism and Western medicine's reductionism. J Anhui TCM Coll 1999;18(1):1–3.

[3] Fan Z. Relationship between modern medical model and the overall concept. Basic Med Educ 2000;2(3):262.

[4] Cheng W. Holism characteristics of Chinese Medicine. J Basic Chin Med 2001;7(6):3–4.

[5] Kuang T. Test development trend of 21st century Western medicine. J Zhejiang Chin Med Coll 2002;26(6):1–4.

[6] Zhao X. Holism in Western medicine. J Integr Med 2012;22(6):353–4.

[7] Yu L, Zou J. The application in medical practice of medical holism. Occup Health 2007;23(12):1043–4.

[8] Yinchao L, Yihong Z. The change of medical model and the characters of TCM holistic. Li Shizhen Med Materia Med Res 2008;19(2):480–1.

[9] Fan S, Hu J, Cao Z. Brief discussion of holism in Chinese medicine. Li Shizhen Med Materia Medica Res 2006;17(10):2057.

[10] Yang Y, Xie L. The differences of holism in traditional Chinese medicine and Western medicine and systemic concept in traditional Chinese medicine. Mod J Integr Tradit Chin West Med 2008;17(22):3426–9.

[11] Yongqiang W. TCM holistic medicine analysis. Liaoning J Tradit Chin Med 2004;31(1): 24–5.

[12] Hu D, Liu M. Holistic medicine – medicine development trend. Chin J Med Guide 2007;9(2):166.

[13] Geng Q. The influence of holistic medicine on the future medical development. Mod Hosp 2010;6(9).

[14] Zhao X. Holism in Western medicine. Shenzhen J Integr Tradit Chin West Med 2012; 22(6):353–4.

[15] Yu L, Zou J. The application of holism in practice medicine. Occup Health (Auckl) 2007;23(12):1043.

[16] Wang Y, Chen Y. Analysis of holistic view in TCM. Liaoning J Tradit Chin Med 2004; 31(1):24–5.

[17] Li Y, Zhao Y. The change of medical model and the characters of TCM holisim. Li Shizhen Med Materia Med Res 2008;19(2):480–1.

[18] Yang Y, Xie Y. Differences between Chinese and Western medicine and TCM holism. Mod J Integr Tradit Chin West Med 2008;17(22):3426–9.

[19] Fan S, Hu J, Cao Z. The holism concept of traditional Chinese medicine philosophy. Li Shizhen Med Materia Med Res 2006;17(10):2057.

[20] Hu D, Liu M. Holistic medicine – medicine development trend. Chin J Med Guide 2007;9(2):166.

[21] Quanguo G, Nan G. Holism of TCM on sub-health. ACMP1 2007;135(15).

[22] Daqing Z. Integral theory in Western medicine. Med Philos 2010;31(2):398.

[23] Yunxi J, Fusheng Z. Micro-holism theory in Chinese medicine. J Chin Med 2010;25(5).
[24] Chen W. About traditional Chinese medicine holism characteristics. Chin J Basic Med TCM 2001;6:3–4.
[25] Hui L, Ma D, Ji Q. Study on relationship between TCM holistic view with holographic biology. J Ninxia Med Coll 2001;3:221–3.
[26] Fan Z. Discussion of modern medical model and holism. Shanxi Med Univ (Preclinical Med Educ Ed) 2000;3:262.
[27] Wu C. GP's holistic medical view. Chin J Gen Pract 2002;2003(3):45–7.
[28] Zhang J, Cai B. Is the TCM holistic view the feature and advantage of Chinese medicine? Med Philos 2002;11:58–9.
[29] Gao Q, Gao N. Chinese medicine total perspective and sub-healthy. Chin Med J 2007;5:7–8.
[30] Yu L, Zou J. The application in medical practice of medical holism. Occup Health 2007;23(12):1043–4.
[31] Yinchao L, Yihong Z. The change of medical model and the characters of TCM holistic. Li Shizhen Med Materia Med Res 2008;19(2):480–1.
[32] Fan S, Hu J, Cao Z. Brief discussion of holism in Chinese medicine. Lishizhen Med Materia Medica Res 2006;17(10):2057.
[33] Yang Y, Xie L. The differences of holism in traditional Chinese medicine and Western medicine and systemic concept in traditional chinese medicine. Mod J Integr Tradit Chin West Med 2008;17(22):3426–9.
[34] Yongqiang W. TCM holistic medicine analysis. Liaoning J Tradit Chin Med 2004; 31(1):24–5.
[35] Liu P. Chinese traditional health maintenance inquiry. Boxing-Martial Art Sci 2013; 10(1):1004–5643 (2013) 01-0082-03
[36] Fan Z. Relationship between modern medical model and holism. J Shanxi Med Univ (Preclinical Med Educ Ed) 2000;vol 3:054.
[37] Kuang T. On development of Western medicine in 21st century. J Zhejiang Univ Tradit Chin Med 2002;6(1).
[38] Xie L. Discussion of the exemplification of TCM whole concept in TCM emergency. Clin J Chin Med 2011;3(17):98–9.
[39] Premal S, Deborah M. The medical model is dead – long live the medical model. Br J Psychiatry 2007;191:375377. doi:10.1192/bjp.bp.07.037242.

Chapter 2

Medical Empathy

Chapter Objectives

After studying this chapter you should be able to:

1. Define empathy
2. Explain the scientific instruments which measure empathy, and how the results are used in medical education
3. Describe how evidence-based medicine and research results and conclusions influence the medical decisions in both Chinese medicine and Western medicine
4. Discuss how practitioner education and competency influences clinical judgment in both Chinese medicine and Western medicine
5. Explain the team efforts between practitioners of Chinese medicine and Western medicine with patients, and how they can affect outcome positively
6. Summarize the context for medical empathy in personalized medicine

OVERVIEW

This chapter describes the qualities of medical empathy. The goal is to positively influence delivery of personalized care in the integrative cardiovascular Chinese medicine setting. The main factor within this goal involves acceptance. To provide or facilitate appropriate care for a patient and cooperate with a health care team, integrative medicine providers should establish and develop certain qualities within themselves which encourage positive outcomes.

The virtuous quality encourages the health practitioner to identify and establish values and beliefs. Through this they can share these values and beliefs with the patient, listen to the patient's own values and beliefs, and determine whether the patient's needs can be supported within the clinical setting. Practitioners are least likely to resort to paternalistic methods in every clinical case; they are more likely to offer their services in a way that helps patients make better health decisions. The intention is for practitioners to accept and respect of the autonomy of patients.

Integrative Cardiovascular Chinese Medicine. DOI: http://dx.doi.org/10.1016/B978-0-12-420014-2.00002-8

The knowledgeable quality encourages the practitioner to expand his or her breadth of medical information, skills and behaviors. Realizing the extent and magnitude of education within other medical systems – their philosophies, clinical intentions, principles and procedures – has potential for acceptance and respect for what other health practitioners do. Expansion can provide confidence to gather additional tools for examination and treatment, or appropriately refer out.

The understanding quality encourages the health practitioner to engage in appropriate interpersonal communication with patients. Practitioners should proceed with concern, compassion, patience and a sense of duty; they are able to understand different medical and health care systems and assist appropriately with the patient's health care team when necessary. Debates over whether empathy is intrinsic or a learned skill have led to the development of special education on methods in clinical communication. The therapeutic relationship should be based on mutual trust, respect and support. Initially, time must be invested to engage, empathize, educate and enlist the patient based on shared values and understanding.

The foundation of empathy is, first, to do no harm. The main intention is to improve patient outcomes by honoring their beliefs and values and using them to personalize their care. A formula for this includes mutual acceptance and respect for the patient and their health care team, learning about different medical systems, and utilizing appropriately their objective and subjective clinical procedures, protocols, algorithms, differentiations, drugs, medicines and other remedies.

Key Points

- How health practitioners view clinical empathy will impact practice and health outcomes.
- Scientific instruments which measure empathy give a clear understanding that empathy is likely to be either intrinsic or learned.
- The preference to practice medicine holistically is an indication towards the desire to show empathy.
- Relational versatility is considered when the patient perceives empathy only in its absence.
- Evidence-based medicine and the scientific methods are used to draw conclusions. These are helpful clues for both Chinese medicine and Western medicine towards developing personalized medicine.
- Cardiovascular residual risk is an enormous challenge in patient management.
- To provide better personalized care options, health practitioners in both Chinese medicine and Western medicine need to establish competency in both systems of medicine.

- Reductionist and heuristic thinking is shared by both Chinese medicine and Western medicine practitioners alike.
- For the purposes of teaming Chinese medicine and Western medicine, the former has specific benefits during the prevention and early-to-upper middle stages of disease progression, while the latter has benefits for upper middle and advanced to end-stage progression.
- Effective patient interviewing methods can provide narrative knowledge and logico-scientific knowledge, both of which contribute to the healing process.

VIRTUOUS QUALITIES

Empathy is a controversial subject with differing opinions about whether it is an intrinsic attribute or can be taught and utilized effectively. Scientists consider it to be a cognitive trait in a well-adjusted person – who therefore has the ability to positively perceive and accept the emotions and experiences of other people. The expectation is that they resolve to treat others as they would want to be treated.

Health care practitioners should develop this trait by taking, assessing and adjusting their own life habits. They should consider dietary and exercise routines, inward reflection, meditation and other forms of balancing physical and emotional wellbeing. Realizing the efforts one must go through to make a positive impact gives a health practitioner great insight. It allows them to improve their verbal and non-verbal communications efforts. They can empathize with difficult patients, connect with people of diverse backgrounds and values, be more forgiving, and objectively guide even a subjective course of treatment. Relational versatility is an operational method in interpersonal communication. Health practitioners can learn how to use this in professional conduct. For many, it assists effective active listening skills. When patients feel that their narrative will be heard and understood it strengthens the practitioner–patient relationship and improves outcomes.

Measuring and Teaching

Medical empathy has become such an important matter in the clinical setting, that investigating how to instill and improve incidences among students and practitioners has become an important part of medical education. Researchers consider empathy to be a concept of variance which can be measured psychometrically. There are many research instruments according to the population being observed. For example, the Interpersonal Reactivity Index, Hogan Empathy Scale and Jefferson Scale of Physician Empathy are a well-known few. The Jefferson Scale of Physician Empathy has been used since the early 1990s to measure attributes of empathy in practitioners – such as competence in history taking, moral reasoning, dutifulness, behavior towards patients, compassion, and standing in the patient's shoes. It is most interesting that women tend to score higher ratings on empathy than do men; it is suggested that the reasons for this have to do with women's natural instincts towards nurturing, and their practice decisions to take in fewer patients and to focus more on prevention and developing their practitioner–patient relationships.

Especially in Western medicine, these findings could have an important impact on the selection of students and residents, counseling on specialty, and strengthening skills and behaviors as practitioners. In Chinese medicine, it could identify ideal students for a career in holistic medicine/health care, selection of students for advanced training or doctoral scholarship, encourage some students who should appropriately specialize, specify practitioner attributes and practices which are favorable to the public perception of the profession, and influence the development of stronger continuing education trends.

Western medicine educational institutions now consider that empathy must be modeled through the behaviors of educators in order for students to recognize and pattern it. Studies have concluded that empathy tends to be higher during the first 3 years of education, declines during clerkship rotations, and is nearly non-existent during residency. At this time there are no known studies that measure empathy among Chinese medicine practitioners before, during and after clinical studies. It would be interesting to observe how the study of basic theories, differentiation of syndromes, and treatment principles influences the perspective of future practitioners towards themselves and their patients. It would be even more interesting to compare those findings with those of Western medicine. Conclusions could be used to improve teaching models which encourage peer support and self-care.

KNOWLEDGEABLE QUALITIES

Evidence-Based Medicine

Evidence-based medicine has been around for millennia, and has shaped all systems of medicine practiced today. Through trial and error, standardizations had to be established in TCM and Western medicine to form the basics in each system. For example, algorithms, differential diagnoses for cardiovascular symptoms and differentiation of syndromes became standards used in clinical practice today. Over time they have been adapted to adjust to shifts and trends in the general population. A byproduct of forming medical systems includes the static indoctrination of the health practitioner. Research design and the details involved in studies often reflect the idiosyncrasies of the indoctrination.

Evidence-based medicine should help the health care practitioner to realize trends and changes in public health and to adapt to them. Research studies serve as proof which assists in making crucial decisions in clinical practice. While evaluating human studies on cardiovascular diseases, practitioners are often concerned with the results and conclusions, but should also consider the value of different variables in the data. The world is continuously interconnected and becoming more geographically diverse. As more populations now have a similar incidence of cardiovascular disease, old questions are now being reviewed along with new questions. When a research study is being designed, are people of different blood types, genetics markers, cultures, races, ethnicities, religions, sexual orientations, beliefs and preferences represented? Can a target group fairly represent the standard for everyone? During the history of cardiovascular research, studies were conducted mostly on middle-aged white males. Their symptom patterns and responses to treatment and recovery became the normal means of measuring everyone with cardiovascular disease. What is different about their responses to medicine compared to those of others? For example, symptom patterns in female subjects have recently been identified as different from those in males. What is incidentally distinct about them from the main target group? How does the incidence of cardiovascular disease in subgroups compare or contrast with the main target group? What factors can be considered the most significant evidence? For example, cardiovascular diseases are the major causes of death in minority populations in the United States. Are genetics, obesity, diet and lifestyle the only concluding factors involved in the causes of cardiovascular diseases? Does any group, including the main target group, really measure up to the results and conclusions of research studies? What is the value of the quality of the corroborating evidence from other studies which proposes to support conclusions? Are objective results and conclusions the only way to determine the likelihood that a treatment will work on an individual patient in the clinical setting?

Treatment and Management

Cardiovascular diseases can become quite complex, and therefore pose a challenge for the practitioner and patient. For example, elderly patients are challenging to treat because of the management and adjustment of optimal blood pressure, and the consequences for the central nervous system of maintaining it too low. Another example involves the problems with considering blood pressure as a target or guide in treatment. It can be inconclusive whether achieving a balance here will be of much therapeutic significance. Health practitioners are faced with the challenges of underlying cardiovascular disease or complications despite control of blood pressure. In addition, there are the different effects and risks associated with complex mixes of medicines, whether they are pharmaceutical or herbal. Finally, there is the patient's compliance to lifestyle adjustment and to the therapeutic regimens suggested.

While evaluating the use of TCM and/or Western medicine in treatment, one approach potentially has a greater clinical value than the other. For example, prevention steps involving modification of lifestyle and recovery from illness or cardiovascular deficits may prompt the use of Chinese medicine, while more aggressive therapies for uncontrolled cardiovascular disease may necessitate the use of Western medicine drugs and/or surgery. Western medicine practitioners involving TCM practitioners in patient management should learn, or at least understand, basic theories, diagnosis, treatment principles and the forms of treatment which may be offered. TCM practitioners providing care for cardiovascular disease patients must establish the necessary knowledge and understanding of subjects in Western medicine. Some subjects include internal medicine, anatomy and pathophysiology of the cardiovascular system. Clinical knowledge and practical skills include mining information during patient interviewing, physical examination, interpretation of laboratory test results, and preparation for factors in practice which involves:

A. *Risk factors.* Practitioners must keep up to date on scientific information, epidemiologic trends and residual risk solutions concerning disease development despite blood pressure stabilization.
B. *Clinical management.* Assessment of cardiovascular disease progression by using routine stress tests and observing vascularization of the retina, evaluating kidney function damage, using serum creatinine, urinalysis to detect albumin, ultrasound to detect and monitor the carotid arteries, and echocardiograms to evaluate left ventricular chamber and tissue dysfunction.

The basics necessary in Chinese medicine include theories covering yin and yang with the viscera and five elements, four stages, six levels, eight principles, meridian channel theory, essence, qi, blood and body fluids. Clinical knowledge includes diagnostic methods of tongue and pulse observation, four diagnostic methods: interviewing, smelling, auscultations and touching, and prevention and treatment principles which cover constitution, etiology and the pathomechanism of cardiovascular diseases in Chinese medicine.

UNDERSTANDING QUALITIES

Clinical Practice Guidelines

The importance of communication between a practitioner and patient involves an understanding of the mutual hopes, fears and preferences. It also involves the desire to realize that the patient narrative reports the individual complexities in his or her life – which influence the state of health. Here, the patient can also present and establish their role in compliance, and practitioners can discuss residual risk and other concerns that direct a rationale for personalized care. Example situations include the resistance to complying with pharmaceutical or herbal medications, a patient's ethnicity, other genetic inheritance or gender which has a historical predisposition to cardiovascular diseases, patients who may require less aggressive treatment such as in pre-hypertensive patients and some elderly patients.

Practitioner Competency

Cardiovascular specialists in Western medicine have some challenges in the communication and referral process with primary care physicians and other practitioners. With the increasing preferences towards including Chinese medicine into the health regimen of patients, collaboration of medical systems becomes even more crucial. The cardiovascular Chinese medicine specialist may either join a medical team in a cardiovascular clinic or mutually refer in private practice. Nevertheless, the collaboration of knowledge and skills guide clinical competency, involving conducting a risk assessment, monitoring patients, encouraging patient compliance, and designing educational efforts for the public and for other health practitioners.

Cardiovascular Chinese medicine specialists collaborating in a multidisciplinary setting, or working in private practice, would likely face some of the same challenges in the referral process as does the Western medicine cardiologist, and may likely view these challenges similarly. As it stands, the primary care physician is the sorting gate-keeper in medicine. They endeavor to provide care at the point of need and refer the patient to a specialist. However, some complaints from specialists about the clinical judgment of many primary care physicians lead to a feeling of mistrust between the medical practitioners – and the patient falls in the middle of the battle.

Major problems with the referral process interfere with the continuum of care. For starters, there are often inconsistencies in the patient records about the cardiovascular risk evaluation, physical assessment, and proper diagnosis. Next is the presentation of extreme types of patient. For example, the first type is the patient with perceived pre-hypertensive symptoms who can self-manage with lifestyle modification and mental/emotional adjustment. The second type is the patient in such a deteriorated end-stage that there may be little that can be done to reverse the health outcome.

Clinical Judgment

When a practitioner becomes accustomed to treating in a certain way it makes them feel satisfied with their abilities to provide similar results in patients. In addition, satisfaction in the results of established methods influences a bias against others, which is called heurism. Having a bias interferes with the discourse between the health practitioner and patient. Limitations in the practitioner's true abilities will become apparent, many patient cues will likely be missed, important information in the narrative may be ignored, and the outcome may not be as favorable.

The culture of the practitioner–patient relationship is considered to be based on reciprocity. For the relationship to function it must include shared values and experiences. The patient can feel more understood by the practitioner, see their own role in the therapeutic process, and therefore accept personal responsibility. For example, in Chinese medicine Liver Qi stagnation, Liver Wind and Liver Fire are common occurrences in everybody (Chapter 6). These internal syndromes are involved in the progression of many diseases, but specifically cardiovascular symptoms and diseases as well. Internal syndromes can be noticed outwardly in physical appearance and emotional behavior, and are most difficult to control by the patients themselves. Practitioners are people too. Even they will expect to have personal experiences with these syndromes, which are often the result of an initial breakdown and continued failure to cope with everyday life situations. The work is demanding, there are family obligations, and relationships with others bring frustration and pleasure. Endeavors and pursuits in life bring many challenges which the practitioner should have personal experience with as well. However, with experience comes wisdom, so the practitioner should have battled with, and achieved, a mastery over these pathologic syndromes. They should have the wisdom to know when to head them off before they have a chance to become the normal life function, and counsel on prevention measures to discourage recurrence. While listening to the patient narrative, and conducting the physical examination, practitioners must know how to detect adverse cardiovascular signs and how to guide the patient to recover from the effects before a certain point. For a holistic practitioner, clinical judgment is not limited to what is done beside the hospital bed or examination table. They should also have learned valuable life lessons which help with transitions a patient must undergo during treatment.

A Team Effort

Chinese medicine and Western medicine practitioners and researchers view evidence of the efficacy of perspectives, procedures, treatments and medicines as important aspects of care. Western medicine practices in research tend to focus on rigorous testing methods in an attempt to recreate consistently similar future results. In the world of Chinese medicine, evidence-based medicine is akin to the big tenacious elephant in the room. It is obvious that it needs acknowledgment; however, the initial reaction has been generally negative. Chinese medicine research perspectives largely tend to reject the principles of the scientific method, or seek to make up excuses for why it should. Efforts focus on defending and reinforcing compulsory ancient methods. On the other hand, testing of the main constituents and efficacy of herbs, substances and medicines seems to follow rigorous international scientific standards. The main idea of research is to offer information about results and conclusions, which provide a means for clinicians to manage care for their patients. The evidence should be in the context of patient needs.

The basic five-element system in Chinese medicine concurs with Western medicine's basic pathology. Both systems agree on the importance of prevention during treatment, considering the collateral damage to other organs during cardiovascular disease progression. Here is where the cardiovascular

Chinese medicine practitioner and the Western medicine cardiologist or primary care physician meet as a team.

It can be necessary to establish guidelines for making clinical judgment while integrating systems of medicine. There are indications and limitations on the perspectives and tools to be utilized. Chinese medicine methods are better for prevention, maintenance, and at times recovery in early to upper moderate stages, while Western medicine methods are better for uncontrolled upper moderate, advanced and end stages.

Western medicine practitioners, such as primary care physicians and cardiology specialists, often have little concern about treating early to moderate hypertension. There are many factors involved in blood pressure fluctuations in the course of a day or with regular living, so it is likely impossible to determine early on whether the changes indicate disease development. To the patient, they may simply suggest lifestyle and mental adjustments as favorable preventive measures. Here is where Chinese medicine can consider the patient based on different criteria, including body constitution, examine the state of balance between the viscera, determine syndrome differentiation, and make more personalized prevention and maintenance recommendations (Chapter 6).

Western medicine practitioners are very concerned about the development of a cardiovascular disease, residual risk, and how aggressive treatment impacts the quality of life. It has been established that educating patients about risk, and controlling or avoiding risk factors, puts the responsibility on them to participate in the team effort for more favorable outcomes.

THE PATIENT NARRATIVE

One of the most important things a health practitioner can do for a patient is to actively listen to them. In many cases, important information reported about the most recent or current state of a person's health is spontaneously presented. Effective listening involves displaying certain attitudes and behaviors, and picking up on non-verbal cues. Aside from the results of diagnostic testing and physical examination, the patient interview is a unique chance to truly understand them. Interviewing should not be conducted like an interrogation of open-ended questions, rather it should invite the patient to contribute information which will help the practitioner to understand how to guide them. Researchers call the health practitioner's preconceptions, along with behaviors and skills used to gather information from the patient interview as either logico-scientific or narrative knowledge. In the logico-scientific, the practitioner is detached, considers the events as general and which illicit universal behavior and emotional responses. In narrative knowledge the practitioner uses cognitive, symbolic and affective methods to listen for and construct the information for understanding the significance of the narrative. Optimistically, the patient narrative contributes to important clues which can influence the therapeutic process.

SUMMARY IN THE CONTEXT OF MEDICAL EMPATHY FOR PERSONALIZING MEDICINE

For providing personalized medicine, it is important to establish the culture of the practitioner–patient relationship. While having mutual experiences in common is an easier way to empathize with patients, scientific methods can be employed to reach the patient as well. In any case, effective listening during patient interviews can properly direct further analysis, and having empathy for the patient opens understanding and trust. The holistic practitioner has the responsibility to educate themselves on how to tailor treatments for specific patient needs. Western medicine and Chinese medicine can partner together and exchange theories, sciences, examination, diagnosis and treatment methods for mutual understanding. Consequently, assuming proper roles and predicting realistic expectations in prevention, maintenance, disease progression and recovery has potential for importance.

FURTHER READING

[1] Sun Y. Reflections on practicing the hippocratic oath with five percepts and ten needs of physicians. Med Philos 2012;33(11A):25–7.

[2] Zhang Y, Li M. Traditional Chinese medicine culture function of ethical construction in the doctor–patient relationship. Hunan J Tradit Chin Med 2011;27(5):103–5.

[3] Yin Y. The inspiration of the traditional medical moral concept to cultivate the modern Chinese medicine clinical practice ability. J Nanjing Univ TCM (Soc Sci) 2012;13(4):205–8.

[4] Zhang C. The reflection of current conflict of medical culture between doctor and patient by the concept of He. ACTA Univ Med NanJing (Soc Sci) 2012;51:278–81.

[5] Jiang J, Zheng S, Li Y, et al. The analysis on the current situation and suggestion of the empirical research about the doctor and patient relationship in China. Med Philos 2012;33(10A):18–20, 71.

[6] Lulu S, Chu C. The current doctor–patient relationship in China. J Chifeng Univ 2012;28(8):95–6.

[7] Li Z, Zhao L. Problems of doctor–patient communication in college students of traditional Chinese medicine and countermeasures. China Foreign Med Treat 2013;1:142–142.

[8] Huang W. Problems of doctor–patient communication in China and strategy. J Hebei United Univ (Health Sci) 2012;15(2):267–8.

[9] Xie Z. Building harmonious doctor-patient relationship in Chinese medicine hospital. Chin Med Mod Distance Educ China 2012;10(24):133–4.

[10] Gui Y, Gui J, Hua R, The guide of history research of humanity of TCM for the clinical medicine [j]. Chin Med Mod Distance Edu China 2011;02:30–1.

[11] Wang M. Trinity study on the plight of doctor–patient relationship [J]. Chin Health Serv Manage 2013;02:97–9.

[12] Mohammedreza H, et al. Physician empathy: definition, components, measurement, and relationship to gender and specialty. Am J Psychiatry 2002;159:1563–9.

[13] Makoul G, et al. Essential elements of communication in medical encounters: the Kalamazoo consensus statement. Acad Med 2001;76:390–3.

[14] Braddock CH, Edwards KA, Hasenberg NM, et al. Informed decision making in outpatient practice: time to get back to Basics. JAMA 1999;282(24):2313–20.
[15] Zoppi K, Epstein RM. Is communication a skill? Communication behaviors and being in relation. Family Med 2002;34(5):319–24.
[16] Suchman AL, Markakis K, Beckman HB, Frankel RA. A model of empathic communication in the medical interview. JAMA 1997;277(8):678–82.
[17] Eide H, Frankel R, Haaversen AC, et al. Listening for feelings: identifying and coding empathic and potential empathic opportunities in medical dialogues. Patient Edu Couns 2004;54(3):291–7.
[18] Hariri AR, Bookheimer SY, Mazziotta JC. Modulating emotional responses: effects of a neocortical network on the limbic system. Neuroreport 2000;11(1):43–8.
[19] Lieberman MD. Social cognitive neuroscience: a review of core processes. Annu Rev Psychol 2007;58:259–89.
[20] Finset A, Mjaaland T. The medical consultation viewed as a value chain: a neurobehavioral approach. Patient Edu Couns 2009;74(3):323–30.
[21] Schulte-Ruther M, Markowitsch HJ, Fink GR, Piefke M. Mirror neuron and theory of mind mechanisms involved in face-to-face interactions: a functional magnetic resonance imaging approach to empathy. J Cogn Neurosci 2007;19(8):1354–72.

Chapter 3

Patient Perspectives

Chapter Objectives

After studying this chapter you should be able to:

1. Explain what should be included in the patient narrative
2. Describe how the constitutional theory is used in Chinese medicine
3. Discuss how the constitutional theory can be used as a health model for cardiovascular patients
4. Discuss treatment and the recovery effort to which patients must contribute
5. Summarize the context of the patient role in personalizing medicine

OVERVIEW

This chapter describes the patient-centered focus within integrative cardiovascular Chinese medicine. In the clinical setting, patients expect the practitioner to listen, understand and provide effective solutions for them. Health practitioners should then expect the patient to be more responsible and involved in the treatment process. To strengthen therapeutic relationships and to improve outcomes, patients and health care providers should focus on three accomplishments which contribute to the success of personalizing medicine.

In the first accomplishment, the ideal health care practitioner establishes partnerships based on mutual respect and trust. They seek to understand the patient on a personal level by incorporating their preferences and beliefs. They guide the patient during the therapeutic process. Cardiovascular disease patients under the care of Western medicine physicians are known to openly or secretly include alternative medicine practitioners on their team. Patients are not always qualified to determine the proper roles or even the appropriateness of different health care methods for their situation. When Chinese medicine is included in cardiovascular care it has a very distinct role. Patients and Western medicine health care providers should understand the indications, contraindications and limitations.

Integrative Cardiovascular Chinese Medicine. DOI: http://dx.doi.org/10.1016/B978-0-12-420014-2.00003-X

In the second accomplishment, Chinese medicine practitioners successfully talk to other practitioners about what they can do for patients, and the realistic expected outcomes. It is also important for health practitioners over all disciplines, who are frequented by cardiovascular patients, to be educated about disease conditions as well as other systems of medicine. To begin with, Chinese medicine practitioners can do more to increase their knowledge about the biomedicine perspective of the cardiovascular system structures and pathophysiologic mechanisms. Further, stronger efforts to increase public and interdisciplinary recognition must be implemented. TCM practitioners must be prepared to describe constitutional theory, syndrome differentiation, treatment principles and herbal medicine selections from the grounded and practical side rather than just lingering on the ethereal and conceptual. The patient can benefit more when practitioners in their health care team can understand terminology and the approaches to care of the other systems, and possibly accept and synchronize them.

In the third accomplishment, patients organize themselves for clinical visits by preparing a clear and detailed narrative which gives insight into habits, beliefs, preferences, and possible factors which cause disease conditions. The cardiovascular patient being treated with Chinese medicine will greatly benefit by incorporating the analysis from constitutional theory, as it is often used to help identity a patient type, to diagnose, and to personalize a treatment plan, and it can be used for the selection of medicines and other remedies.

Key Points

- Patient-centered care focuses on the needs of patients by respecting their preferences and beliefs.
- Patients must also establish an understanding about their emotions, diet, lifestyle, treatment preferences and beliefs in a verbal or written report called a narrative. This report is disclosed to the health practitioner.
- The constitutional theory in Chinese medicine is used by practitioners to identity a patient type and for diagnosing, creating a treatment plan, and selecting medicines and other remedies.
- Cardiovascular patients can use the constitutional theory to establish a health belief with preferences which can assist with diet and lifestyle modifications, diagnosing and treatment selection.
- The systems of Chinese and Western medicine have a distinct role, indications, contraindications and limitations in the treatment process for cardiovascular patients.
- Chinese medicine theories and procedures at times overlap Western medical sciences and practice, and impact disease states as well.
- Health practitioners should be well educated about other systems of medicine, especially since cardiovascular patients often gather their own health teams.

ISSUES IN THE PATIENT–HEALTH PRACTITIONER RELATIONSHIP

In patient-centered care, the practitioner establishes the mutual trust in the therapeutic relationship by understanding and respecting the patient on a personal level. The key disparities where understanding and respect are especially critical are where it concerns the patient's race, culture, nationality, religion, sexuality, gender, age, weight, occupation, social classification, personal values and preferences. The positive role of the health practitioner must be maintained to provide the best clinical outcome.

Practitioner Bias

It is well documented that health practitioners without much empathy for a demographic contribute to less-informed patients with limited involvement in the treatment progress. Health care practitioners who are considered insensitive and bigoted often make broad generalizations and assumptions. They often verbally dominate, interrupt, and misinterpret during patient narrative reporting, and misunderstand symptoms.

For example, in the United States it is known that, among all racial and ethnic groups, African Americans lead in the incidence of cardiovascular disease and death from complications. Native Americans, Hispanics and Asians also have a high incidence as well. Cardiovascular disease symptoms have been historically ignored in women. This was due to guidelines in health monitoring being patterned from symptoms in men. Some other groups and cases where there are some disparities involve patients with undetermined or underdeveloped symptoms. In addition, many published research studies have concluded that the health practitioner's racial discrimination, cultural insensitivity, gender and sexual biases, and language barriers, have been some main factors involved in contributing to poor-quality outcomes.

THE BODY'S CONSTITUTIONAL SYSTEM

The constitutional theory in Chinese medicine was adapted from the original five-element theory thousands of years ago. Today it can be used as a personalized medicine approach according to a person's body constitution. In a basic sense it describes the ordinary, pathologic and recovery states of health according to the transitions of the four seasons of the year. A person can view diet and lifestyle suggestions during each of the four seasons, and consider how to adapt and optimize their health to the season or primary climate where they currently live. In addition, the body's constitution system explains the ordinary human characteristics according to stages in life. Disease prevention, pathologic factors and recovery methods are evaluated according to the Chinese medicine differentiation of syndromes. The role of the constitution system in cardiology is to provide patients with more health options, empower them with personal responsibility in the prevention, treatment and recovery process, and also to use a modifiable and personalizable method of care between the prevention and serious stages.

The patient following this system in a basic sense should realize how their basic constitution according to gender, and life milestones, has an impact on emotional, dietary and lifestyle choices for disease prevention, treatment and recovery. Consequently, doing so can allow them to be more engaged in decision making, even if the practitioner takes a more dominant role in the relationship.

For the health practitioner it not only offers empowerment for patient personal accountability, it also has the potential for the practitioner to consider and modify their own clinical approaches. By following the five-element seasonal pattern with the patient, practitioners might be more willing to demonstrate more empathy and acceptance. The patient may be more willing to participate in the therapeutic process by coordinating their mental and emotional feelings, and their diet and lifestyle, with the requirements for preventing cardiovascular disease or for promoting recovery from such disease.

Balance of the Constitution

The origin of the body constitution comes from the innate factors of inheritance from the parents and postnatal factors involving nutrition and lifestyle. Constitutional theory is an original virtue in Chinese medicine which is based on five-element theory (Chapter 1) and uses it to describe the balance and imbalance of the body by describing adaptation to the environment, contour shape, physical features, emotions, intelligence, spirituality and pathologic states. The body constitution must be maintained as close as possible to the ideal state for good health. The mental state and lifestyle are important as the essence, yin, yang, qi blood and body fluid balance is so closely connected to their states. Irregularities in the diet and lifestyle will cause more adaptation to negative stressors, and creates a situation in which the body is forced to take longer or more effort to regain balance. Prevention of disease is part of maintaining good health. Seasonal care can be part of a strategy in a prevention program. The basic principle is to cultivate the yang in the spring and summer and the yin in the fall and winter.

Prevention: General Suggestions Based on the Four Seasons (Table 3.1)

Constitution: Basic Types based on Gender and Milestone (Table 3.2)

Lifestyle: Disease Prevention based on Gender and Milestone (Table 3.3)

TABLE 3.1

	Basic Seasonal Habits for Cardiovascular Disease Prevention
Spring	**Diet:** cultivate the yang and regulate the smooth flow of liver qi. Prevent adverse influence on the blood pressure by blending more sour flavors into daily meals. Include sour herbs which activate yang and immune mechanisms of wei qi to prepare for the autumn.
	Exercise: include vigorous activities, jogging, tai qi which stretch the muscles and tendons and increase energy, can be done outdoors or indoors due to the weather.
Summer	**Diet:** cultivate the yang and regulate the rising of heart yang. Balance bitter, salty, acrid and sour foods in moderation in daily meals. Eat vegetable soups with various seasonal fruits and vegetables and herbs which maintain body temperature and blood circulation. Drink water and other fluids in moderation to avoid edema.
	Exercise: moderate activities such as tai qi or sports which stretch the muscles and tendons and maintain energy, can be done outdoors.
Autumn	**Diet:** nourish the yin and regulate the lung qi and spleen qi. Mix and alternate choices of seasonal vegetables and fruits with sour, bitter, pungent flavors with sweet and reducing the acrid flavors.
	Exercise: the accumulated energy of the summer should be stored for the coming winter. Exercise should be moderate to slow. Some activity can be done outside and indoors due to the weather.
Winter	**Diet:** nourish the yin and maintain heart and kidney yang balance. Eat less salty and more bitter and acrid food.
	Exercise: moderate to slow indoor exercises to avoid opening the pores too much and losing yin fluids.

THE ORGANIZED PATIENT

The therapeutic relationship between the cardiovascular patient and the health practitioner involves assuming roles, mutual give and take, and not crossing boundaries. A patient must enter into the relationship prepared to carry out their role fully, otherwise they share the responsibility of poor outcomes. If the practitioner's role is to listen actively to the patient narrative, share insight, experiences, recruit the patient into the partnership and provide therapeutic options, then the patient's role is to provide organized, direct, detailed and honest accounts of their personal and family health history, personal behaviors, thoughts and emotions, as well as dietary habits and lifestyle patterns. The patient must also report their experiences in the therapeutic relationship. It can be helpful to explain what treatment approaches worked, and which did not. Also, if a change in diet and lifestyle is necessary – such

TABLE 3.2

Constitution	General Body Characteristics
Basic male constitutional	**Yang type:** excessive amounts of yang and blood, sexual desires or activities which rapidly depletes essence, yin and blood. The balance of yin and yang through the nourishment of essence and blood is the life challenge of the male.
Basic female constitutional	**Yin type:** excessive amounts of qi, insufficient amounts of yang and blood. Pathological evidences are noticed between the menstruation, childbirth and menopausal stages. Balancing yin and yang through building and nourishment of blood is the life challenge of the female.
General stages of infant and childhood constitution	Yang characteristics of rapid development and growth. Potential for excessive heat and fire syndromes in the liver and heart, deficient immunity wei qi, deficient spleen and kidney syndromes. The personality stage is between agreeable and obstinate.
General stages of teenage and young adult constitution	Continuous rapid growth and development in adolescence which leads to a plateau in adulthood. Yin and yang are balanced, essence generates, qi moves freely and blood flows smoothly. Characteristics include the development of mental and emotional individuality, maturity and self perception, sexuality, morality, virtues and choices. The personality is changeable.
General stages of the middle-age constitution	The developmental plateau leads to decline of vitality. Essence is depleted and there is instability of balance between yin, yang, qi and blood. The personality is stable yet mental and emotional status begins to decline. The viscera, other organs begin to show signs of dysfunction. The menopausal and climacteric stages in both males and females begin.
General geriatric constitutions	The rapid decline of vitality leads to eventual mortality. Essence has long been depleted with near depletion of immunity wei qi, yin, yang, qi and blood. Stagnation of body fluids accumulates along with poor circulation and elimination. Dampness and phlegm is more abundant. The personality is changeable with mental and physical deterioration.

TABLE 3.3

Constitution	General Prevention Suggestions
Basic male constitutional	**Diet:** foods, drinks, herbs and supplements which maintain yin and yang balance. Specifics include tonifying yang and building and nourishing blood and kidney essence.
	Exercise: include activities which moves and nourishes blood, builds and strengthens muscles and tendons, and generates masculine vitality.
Basic female constitutional	**Diet:** foods, drinks, herbs and supplements which maintain yin and yang balance. Specifics include nourishing yin and blood before and after monthly discharge, maintaining mood by smoothing and freeing the flow of liver qi, assisting digestion and water metabolism by tonifying spleen qi and kidney qi.
	Exercise: include activities which moves and nourishes blood for muscles and tendons, skin and hair, circulation of feminine vitality.
General stages of infant and childhood constitution	**Diet:** foods, drinks, herbs and supplements which establishes the balance of yin and yang. Specifics include tonifying kidney essence for intelligence, soothing liver qi and reducing heart fire for colic and uncontrollable behavior problems, and tonifying deficiency spleen qi incontinence and drooling.
	Exercise: activities which promote intelligence, fitness, self awareness and responsibility.
General stages of teenage and young adult constitution	**Diet:** foods, drinks, herbs and supplements which maintains the balance of yin and yang. Specifics include tonifying and regulating kidney essence, soothing and smoothing the liver qi, reducing heart fire, tonifying the kidney essence, lung qi, and spleen qi. The focus is to meet the stressors and demands of puberty, emotional stabilization development of self-expression, irregular dietary habits, career, physical and sexual activities.
	Exercise: activities which regulate energy builds and strengthens muscles and tissues, and regulates fat.
General stages of the middle-age constitution	**Diet:** foods, drinks, herbs and supplements which repairs the imbalance of yin and yang. Specifics include tonifying and regulating qi, nourishing and moving blood, smoothing the flow liver qi, balancing kidney yin and yang, and promoting activity, relaxation and sleep.
	Exercise: activities which promote energy and relieve stress.

(*Continued*)

TABLE 3.3 (Continued)

Constitution	General Prevention Suggestions
General geriatric constitutions	**Diet:** foods, drinks, herbs and supplements which continually work to re-balance yin, yang, qi, blood and body fluids. Specifics include building and protecting immunity and wei qi, circulation of blood, proper movement of body fluids and wastes, draining dampness and phlegm accumulation.
	Exercises: activities which stretch the tendons and muscles and assist with blood and fluid circulation.

as after a diagnosis of atherosclerosis, hypertension, etc. – then the patient should report and explain which preferences and beliefs can be comfortably implemented. Finally, the patient should explain their commitment to incorporating the changes into the therapeutic partnership with the practitioner. Some patients may feel that verbal reporting during health practitioner dictation is more effective, while others may feel that writing and submitting the information during the clinical visit may be better.

Using the Body Constitution Method to Prepare for the Integrative Care Setting

Patients can first use information about the body constitution to estimate general health requirements based on gender and milestone. Next, they can compare their symptoms with the information on pathologic body constitution. This combined information can be used to help the patient develop a realistic self-perspective and to prepare the narrative – which requires the reporting of vital information about the health history, chief symptoms, emotional, dietary and lifestyle habits.

During the interview, the health practitioner can use the time during the patient narrative to develop a deeper understanding. The practitioner – using body constitution theory as a tool to assist in diagnosis and formulating a treatment plan – can interpret the findings and use them to help the patient recover (Chapters 6 and 7).

Integrative cardiovascular Chinese medicine is a health care model that includes personalizing treatments using the patient narrative, blended physical examinations, body constitution theory, differentiation of syndromes, and establishing and modifying a treatment principle and plan. Patients must be able to express the preferences and beliefs about integrating Chinese medicine to improve personal life quality. They should have the knowledge, skills, attitudes, self-awareness and any scientific or practical evidence necessary to partner with all health care providers.

The Patient's Narrative

Before any cardiovascular interview or follow-up office visit, patients should begin or continue the healing process by drafting an effective narrative. The information in the narrative of an initial interview should include, at least, a detailed health history, present complaints and duration, emotional states, dietary, medication regimen and lifestyle habits. The information in the follow-up narratives should include, at least, a summary of health practitioner suggestions (remedies/supplements/medication, dietary and lifestyle modifications), a schedule which followed the suggestions, a status on the symptoms, and a daily log of actual meals and lifestyle habits during the course of treatment.

Next, the patient can decide on the appropriate method of communication for addressing personal health information. For example, some patients prefer to report the information verbally, which has pros and cons. One downside may be the language or cultural barrier – which can be difficult to overcome. Some find that regular entries in a health diary not only help them to keep up with information to report later, it also gives a more thorough look into the unique daily patterns that contribute to health quality. A health diary is often used by patients managing a diet, sleep problems, migraine headaches, weight loss, chest pain or palpitations, menstrual cycles, and other health concerns. Others find that combining the verbal and written aspects on the narrative gives a chance for explanation of points in the journal while the health practitioner writes the report in the chart notes.

Provided that the health practitioner and patient relationship is truly based on empathy, respect and trust, the practitioner can listen and/or read through the narrative, mine through the patient's testimony and hopefully ask nonjudgmental questions. The purpose of questioning is to encourage patients to reflect more on their own motivations towards health improvements. The patient can then determine the steps needed to make changes, and hopefully accept advice and guidance from the health practitioner. Consequently, the patient can perform the indications on a treatment plan and monitor changes in the health status by recording them in the health diary.

Organizing the Interactions between Practitioners in the Health Care Team

Research has revealed that most patients who seek care outside Western medicine are not always looking for a replacement. They may quietly gather a select team of practitioners who can support and serve them as needed. A typical health care team for many cardiovascular patients with moderate symptoms may involve a primary care provider: MD or DO, a chiropractor and a massage therapist for stress relief, an acupuncturist for blood pressure and pain reduction, a naturopath to give guidance on proper nutritional analysis and supplementation or as an alternative physician for various internal medicine diseases, a nutritionist to guide on proper dietary habits, a personal trainer to help with proper exercise techniques, and a psychologic counselor for sorting out the matters of life.

For this reason, it is necessary for health practitioners across systems and disciplines to have shared systems of mutual communication to facilitate care continuity.

Nowadays, patients most likely have had some experience of receiving treatment in Chinese medicine. The most common will include tongue and pulse diagnosis, acupuncture, and herbal medicines. Practice history and published research have shown the efficacy of acupuncture, various herbs, medicines and combination therapies involving varieties of surgical, prescription and OTC Chinese medicines with varieties of surgical, prescription and OTC pharmaceutical Western medicines.

Integrative cardiovascular Chinese medicine blends the systems of Western medicine and Chinese medicine for examination, diagnosis and recommendations for therapies that have the greatest potential for benefit with the least potential for harm. Chinese medicine practitioners who focus on, or specialize in, the care of the cardiovascular patient and Western medicine practitioners such as primary care providers and cardiovascular specialists must form communication bonds to provide better quality options for cardiovascular patients.

The development of integrative cardiovascular Chinese medicine is in recognition of the widespread incidence and prevalence of cardiovascular disease in the United States and worldwide, and serves as a treatment and recovery option for many patient types.

RECOVERY METHODS FOR PATHOLOGIC CONSTITUTION TYPES ACCORDING TO CHINESE MEDICINE

Pathological Body Constitution Types (Chapter 6 Diagnostics and Differentiations)

Pathology: Deficient Constitutions (Table 3.4)

Pathology: Stagnation, Stasis and Excess Constitutions (Table 3.5)

Recovery: Lifestyle Adjustments (Tables 3.6 and 3.7)

Factors in Treatment and Recovery

Integrative cardiovascular Chinese medicine is based on simple and advanced medical sciences, and theories in Chinese medicine. Symptoms, disease progression or recovery is detected and managed through tongue

TABLE 3.4

Constitution	General Pathological Characteristics
Yin deficiency constitution	**Characteristic:** flushed complexion, dry throat, dry skin and hair, emaciated and weak body structure, interior heat sensation and at the center of the chest, feverish palms of hands and foot soles, intolerance for summer weather or tropical regions.
	Behavior: extroverted personality, manic psychoses, hyper sexual desire or activity.
	Causes: deficiency of yang, innate essence insufficiency, premature delivery, excessive sexual activity, illicit recreation and designer drug use, certain prescription drugs, decline in old age.
Yang deficiency constitution	**Characteristic:** pale complexion and lips, obese body structure, muscle atrophy and weakness, cold sensation from the interior and circulating throughout the body. Aversion to cold environments and climates, fluid accumulation especially in limbs and midline, diminished or lack of sexual desire, spontaneous sweating, clear profuse urination and poor or weak defecation, intolerance for winter weather regions.
	Behavior: introverted personality, depression.
	Causes: deficiency of yin, premature delivery, insufficient post-natal care, severe emotional trauma, multiple close duration pregnancies/deliveries, weakness and nutritional and emotional imbalance during pregnancy, prolonged illness, peri-menopause, certain bacterial, viral or fungal infections, decline in age.
Yang deficiency constitution	**Characteristic:** Round chest and/or waistline with cold limbs, edema, shortness of breath and abdominal fullness, fatigue, dull complexion spontaneous sweating, low voice, no desire to speak much, intolerance for hot summer weather and tropical climates.
	Behavior: introversion and emotional instability.
	Causes: deficient innate and acquired essence, anorexia, prolonged or under-treated illness and advancing age.
Blood deficiency constitution	**Characteristic:** emaciated appearance, dizziness, muscle atrophy, numbness and tingling of the limbs, pale complexion, eyes, lips and nails, limp and dull brittle hair and easy hair loss. **Behavior:** introverted or extroverted personality, depression.
	Causes: deficient innate essence inheritance, post-natal malnutrition, prolonged untreated illness, severe bacterial, viral, parasite or fungal infections, nutritional deficiencies, tumors, irregular menstruation, hemorrhage, illicit and designer drug use, aggressive prescription drug therapies.

TABLE 3.5

Constitution	General Pathological Characteristics
Qi stagnation	**Characteristic:** thin body or fluctuating weight gain or a round waistline, intolerance for prolonged heat environments.
	Behavior: introverted personality, mental instability, emotional sensitivity.
	Causes: prolonged emotional and psychological trauma due to fright, grief, disappointment, depression, physical and mental overstrain.
Blood stasis constitution	**Characteristic:** thin body, abnormally patterned pigmented complexion, dark orbital cavity, purple lips or tongue, hair loss, dry skin calluses or squamation, susceptibility for cold, flu and infections, intolerance for windy weather and cold climates.
	Behavior: depressed personality, amnesia, irritability.
	Causes: inhibited blood circulation, retention of blood in the body, fluid deficiency in the vessels, kidney deficiencies or failure, infections.
Damp phlegm constitution	**Characteristic:** obese body, protruding soggy waisteline, heavy body sensation, oily and sallow skin complexion especially the forehead and nose, acne, puffy face, limb edema, irritable personality, intolerance for damp tropical climates.
	Causes: heredity, irregular daily routines, imbalanced diet of sweet and fatty food, insufficient water intake, lack of exercise and irregular sleep schedule.
Damp heat constitution	**Characteristic:** obese body appearance, inability to adapt to a wet tropical environment, oily and dirty or dark yellow complexion and eyes, vexed personality.
	Causes: exposure to rain, diet of sweet and fatty food, abnormal alcoholic beverage consumption habit, prolonged edema, tropical environments, digestive disturbance, summer heat climate which transforms into the autumn, insufficient fluid circulation, bacterial or viral infection, bladder or kidney infections, liver or kidney failure, gall bladder disorders.

and pulse diagnosis as well as physical examinations and diagnostic testing common to Western medicine. Using the body constitution as a therapeutic health model involves a discussion about the constitution, patient self-awareness, commitment to care based on preferences and beliefs discipline.

TABLE 3.6

Constitution	Recovery Suggestions for Stagnation, Stasis and Excess Constitutions
Yin deficiency constitution	**Diet:** foods, drinks, herbs and supplements which nourish yin and blood and subdue yang hyper energy. Avoid spicy, hot, astringent, greasy, alcohol.
	Life: organize and maintain regular daily schedules of waking and sleeping, meals, work and recreation, avoid overstraining efforts in work, sex, avoid hot weather and rooms, tobacco, marijuana products, OTC remedies, prescription and illicit recreational drugs which tend to dry yin fluids.
	Exercise: avoid activities which create a lot of sweating.
Yang deficiency constitution	**Diet:** foods, drinks, herbs and supplements which tonify yang and reduces yin by warming the spleen and kidney yang. Avoid cold, raw, bitter, greasy.
	Life: organize and maintain regular daily schedules of waking and sleeping, meals, work and recreation, avoid cold weather and rooms, tobacco, marijuana products, OTC remedies, prescription and illicit recreational drugs which tend to deplete the proper accumulation of yang while reducing yin fluids.
	Exercise: avoid activities in cold or damp environments.
Qi deficiency constitution	**Diet:** foods, drinks, herbs and supplements which invigorate the spleen, tonifies and nourishes the blood and builds immunity and wei qi. Avoiding greasy, cold, bitter, sweet and spicy.
	Life: organize and maintain regular daily schedules of waking and sleeping, meals, work and recreation, avoid tobacco, marijuana products, OTC remedies, prescription, and illicit recreational drugs which tend to dissipate qi.
	Exercise: avoid continuous repetitious activity.
Blood deficiency constitution	**Diet:** foods, drinks, herbs and supplements which tonifies and nourishes blood flow. Avoid spicy and astringent flavors which vent, dry or deplete yin fluids.
	Life: organize and maintain regular daily schedules and limit study times and other periods of concentrated focus. Avoid tobacco, marijuana, OTC remedies, prescription and illicit recreational drugs which tend to deplete or make the blood dry and apt to stagnation or stasis.
	Exercise: avoid activities which create a lot of brute effort and overstrain.

TABLE 3.7

Constitution	Recovery Suggestions for Stagnation, Stasis and Excess Constitutions
Qi stagnation	**Diet:** foods, drinks, herbs and supplements which promote the regulations of qi of the heart, spleen, lungs and liver. Avoid cold and astringent products.
	Life: organize and maintain a regular daily schedule, avoid tobacco, marijuana, OTC remedies, prescription and illicit recreational drugs which tend to stagnate liver qi.
	Exercise: avoid activities which create a lot of brute effort and mental overstrain.
Blood stasis constitution	**Diet:** foods, drinks, herbs and supplements which invigorates the spleen, promotes movement of qi and moves and circulates blood. Avoid spicy and astringent products which vent, dry or deplete yin fluids, cold and astringent.
	Life: organize and maintain a regular daily schedule, avoid tobacco, marijuana, OTC remedies, prescription and illicit recreational drugs which tend to deplete or make the blood dry and apt to stagnation or stasis.
	Exercise: avoid activities which create a lot of brute effort and overstrain.
Damp phlegm constitution	**Diet:** foods, drinks, herbs and supplements which invigorates the spleen qi and yang, drains dampness and activates the san jiao. This patient is in danger of developing hyperlipidemia and coronary artery disease. Avoid greasy, sweet, cold and astringent flavors which nourish yin fluids or generate phlegm.
	Life: organize and maintain regular daily life of sleep, rest, work and activity. Avoid tobacco, marijuana, OTC remedies, prescription and illicit recreational drugs which tend to contribute to factors of phlegm stagnation and dampness. Make sure the living environment is dry and free of fungus.
	Exercise: avoid activities which create a lot of brute effort and overstraining, do exercises which allow normal movement sequences such as walking, jogging, taiqi, swimming, etc.
Damp heat constitution	**Diet:** foods, drinks, herbs and supplements which invigorates the spleen qi, drains dampness, clears heat and activates the san jiao. This patient may have concurrent kidney, bladder or vaginal infection. Avoid greasy, sweet, hot, spicy and astringent flavors which generate heat.
	Life: organize and maintain a regular daily schedule, avoid tobacco, marijuana, OTC remedies, prescription and illicit recreational drugs which tend to be more astringent, compromise the immune system wei qi, and generating heat in the body.
	Exercise: incorporate activities which require brute effort and vigorous movement to remove the dampness and vent the heat.

Pathology

Patient type 1 has cardiovascular symptoms which have progressed from early to upper moderate; such patients are more likely in the prevention, treatment and recovery category, and they may have the outpatient clinic intervention or exploratory procedures or regular referral visits in private office practices. These patients are also candidates for the benefits of Chinese medicine – alone or integrated with Western medicine – and usually they have the most favorable outcomes.

Patient type 2 has symptoms which have progressed into heart failure and include other complications which usually require closer monitoring, frequent hospitalization and procedures with stenting, ablating, or resections, and may even be selected for surgical procedures during the course of treatment. These patients have poor life outcomes. The use of Chinese medicine alone, or integrated with Western medicine, is an approach restricted to hospital specialists in cardiology who are familiar with the intravenous use of Chinese medicines and Western medicines to rescue a patient on the brink of death.

Some health complications, such as type 1 and 2 diabetes, class III and IV hypertension, cardiomyopathies, heart failure, myocardial infarctions, various bacterial and viral infections, arrhythmias, etc. have cyclical occurrences and are often amplified by seasons, emotions, diet and lifestyle choices.

(a) *Seasonal factors*. Severe weather imposes additional demands on the body. For example, during summer the hot weather causes sweating and loss of fluids, raises blood pressure and can cause syncope. During winter months the cold weather causes difficulty in maintaining body temperature, affects the lungs. In addition it is the season for colds and flus, and other bacterial and viral infections which threaten to dissipate blood volume and quality. Some patients with serious cardiovascular diseases may develop complications or events like a myocardial infarction, which sends them to the emergency room.

(b) *Emotional and behavioral factors*. These are factors which involve romantic relationships, work, career and business relationships, family responsibilities, friendship bonds, meeting of life goals and aspirations. Inability to cope results in depression, psychosis, introversion, extroversion, unhealthy sexual activities, irritability, vexation and amnesia.

(c) *Dietary issues*. These commonly involve water intake, fat intake, sugar intake, nutritional deficiencies, food combining, meal scheduling, and sometimes injury from the novice self-prescribing herbs and nutritional supplements.

(d) *Lifestyle choices*. These are choices which involve career demands, stress levels, exercise habits, sexual activities, family bonds, personal morals and spiritual values, illicit recreational habits which may include tobacco, alcohol and drug abuse.

Treatment and Recovery

During treatments integrating Chinese and Western medicine, changes and transitions often occur – bringing favorable results quickly for some and slowly for others. The reasons are often referred back to the patient contributing their own commitment to recovery. After a course of treatment has begun, and the body recognizes that the motion for balance has been initiated, the patient begins to feel and experience transformations in health which should be documented in the daily health diary.

Recovery is a subjective state. The controversy is in estimating the magnitude and duration. Western medicine and Chinese medicine researchers often debate about the definition of recovery from a holistic point of view. It is agreed that health practitioners partnering with patients to establish early prevention methods provide better outcomes. The consideration of what constitutes recovery is not always according to health practitioner opinions through their systems of treatment. However, respect for the perspectives of each medical system that explains health, disease and recovery, can help influence better medical decisions, and outcomes as well. The physiologic function of the heart and vessels, realizing the demand placed upon them to serve the needs of the entire body, noticing the effect that various pathogenic factors impose upon the blood, the vessels and heart chambers, investigating the sections of the heart and the impact that their failures have upon the integrity of the cardiovascular system.

SUMMARY IN THE CONTEXT OF PATIENT PERSPECTIVES FOR PERSONALIZING MEDICINE

For integrative medicine to provide effective personalized medicine solutions, cardiovascular patients must take responsibility for their own health and actively provide information during the patient narrative. Chinese medicine practitioners have been using the body constitution theory in diagnosing and personalized treatment strategies for hundreds of years. As a health model, the body constitution theory suggests prevention measures which strengthen the body through normal gender and milestone limitations. It also points out pathologic states, which can affect any constitution, and dietary and lifestyle recommendations during treatment and recovery. Patients and health practitioners can use the body constitution method as a tool in the therapeutic relationship. For patients it can provide a guide for empowering them to take responsibility and showing them how and where they have control over the state and outcome of their own health. It can also act as a template to help establish a health identity for the patient narrative which guides the health practitioner to personalize the treatment plan.

FURTHER READING

[1] Guo L, Xu Z, Liao S. Constructing harmonious doctor–patient relationship in public large. Chin Med Ethics 2012;25(6):744–5.

[2] Yugang G. Improving doctor–patient relation to build a harmonious medical environment. Chin Hosp 2012;16(2):57–9.

[3] Wei J. Evaluate trauma surgery achievement in ancient traditional Chinese medicine. Chin J Tradit Med Traumatol Orthopedics 1996;4(6).

[4] Zuo J. On the doctor–patient relationship and countermeasures. Mod Hosp 2010; 10(007):124–5.

[5] Li W. Management of building a harmonious doctor–patient relationship. Clin Ration Drug Use 2013;6(3):160–1.

[6] Zhao L, Chen X, Liu S, Sun B. Present situation of the strained doctor–patient relationship in China and the analysis on its deep reason. Mod Hosp Manag 2013;11(1).

[7] Zhang S. Focus on improving the doctor–patient relationship. J Tradit Chin Med Manag 2013;21(1).

[8] Li Z, Leng M. Doctor–patient relationship research summary. Chin Hosp Manag 2009; 29(3):40–3.

[9] Dan L, Zhi C, Xiang F. Causes of tension between doctor and patient and countermeasures. Contemp Med 2013;19(9):308.

[10] Jian Y. Setting up perfect physician–patient communication to promote a harmonious physician–patient relationship. Chin Med Ethics 2010;23(1).

[11] Jianhua L, Jianguo W, Rui P, et al. China's current sociological study of the doctor–patient relationship. Chin Hosp Manag 2004;24(2).

[12] Zhou Y. Analysis of Chinese ancient psychological treatment. J Jiaozuo Univ 2005; 4(*October*):105–6.

[13] Nina H, Wei C, Li C. Ancient Chinese model of doctor–patient relationship. Chin Med Ethics 2007;21(3).

[14] Mingli S, Yalan F. Study of foreign medical security system under the mode of the doctor–patient relationship. Med Soc 2009;22:41–4.

[15] Stewart M, Brown JB, Donner A, et al. The impact of patient-centered care on patient outcomes. J Fam Pract 2000;49(9):796–804.

[16] Street RL. Analyzing communication in medical consultations: do behavioral measures correspond to patients' perceptions? Med Care 1992;30(11):976–88.

[17] Roter DL. Observations on methodological and measurement challenges in the assessment of communication during medical exchanges. Patient Educ Couns 2003;50(1):17–21.

[18] Mead N, Bower P. Patient-centered consultations and outcomes in primary care: a review of the literature. Patient Educ Couns 2002;48(1):51–61.

[19] De Haes H. Dilemmas in patient centeredness and shared decision making: a case for vulnerability. Patient Educ Couns 2006;62(3):291–8.

[20] Beach MC, Inui T. Relationship-centered care: a constructive reframing. J Gen Intern Med 2006;21(Suppl. 1):S3–8.

[21] Gortner EM, Rude SS, Pennebaker JW. Benefits of expressive writing in lowering rumination and depressive symptoms. Behav Ther 2006;37(3):292–303.

[22] Mezzich JE. Psychiatry for the person: articulating medicine's science and humanism. World Psychiatry 2007;6(2):65–7.

[23] Charon R. The patient–physician relationship. Narrative medicine: a model for empathy, reflection, profession, and trust. JAMA 2001;286(15):1897–902.

[24] Charon R. Narrative and medicine. N Engl J Med 2004;350(9):862–4.
[25] Kim SS, Koplowiz S, Johnston MV. The effects of physician empathy on patient satisfaction and compliance. Eval Health Prof 2004;27(3):237–51.
[26] Adler HM. Toward a bio-psychosocial understanding of the patient–physician relationship: an emerging dialogue. J Gen Intern Med 2007;22(2):280–5.
[27] Karthikeyan G, Pais P. Clinical judgment and evidence-based medicine: time for reconciliation. Indian J Med Residency 2010;132(5):623–6. doi:10.4103/0971-5916.73418 PMCID: PMC3028959.

Chapter 4

Significance of Cardiovascular Symptoms

Chapter objectives

After studying this chapter you should be able to:

1. Describe the Chinese medicine differentiations associated with each symptom of cardiovascular disease
2. Discuss the cardiac and non-cardiac causes of each symptom
3. Discuss the Western medicine diseases which might be diagnosed according to the symptoms
4. List genes associated with certain symptoms
5. Discuss the significance of physical examination procedures used to investigate each symptom

OVERVIEW

This chapter describes the symptoms which may be significant in cardiovascular disorders. The five main symptoms of cardiovascular diseases are chest pain, dyspnea, palpitations, edema and claudication. The main concerns are whether the symptoms are cardiac or non-cardiac in origin.

In Chinese medicine, chest pain is xiong bi. There are two types of chest pain: acute and chronic. In Chinese medicine, dyspnea is considered qi chuan; there are two types: dyspnea and orthopnea. Dyspnea often happens during movement, and orthopnea often happens at rest. In Chinese medicine, heart palpitations are considered xin ji; generally, these are considered as the awareness of beats which have abnormal rhythm characteristics.

In Chinese medicine, syncope is considered xuan yuan; it is characterized by a sudden loss of consciousness. In Chinese medicine, edema is considered

Integrative Cardiovascular Chinese Medicine. DOI: http://dx.doi.org/10.1016/B978-0-12-420014-2.00032-6

shui xhong; it is characterized as excess fluid in the body, and is either localized or peripheral. In Chinese medicine, claudication is in the category of vessel bi; the pain appears when the patient attempts to walk, but it is relieved if the patient stands still for 2 minutes.

Key Points

- Chest pain can be acute or chronic, and it can be due to cardiovascular causes or non-cardiovascular causes.
- Pain, its location, the time when it occurs, and whether physical stress causes it, are important factors when determining whether the cause of chest pain is angina.
- The Hering–Breuer reflex is activated in dyspnea and orthopnea.
- When palpitations are due to arrhythmia, the types of concern in Chinese medicine are paroxysmal atrial fibrillation, sinus tachycardia, and paroxysmal ventricular tachychardia.
- In Chinese medicine, palpitations are xiong bi/chest obstruction and xin ji/palpitation.
- In Chinese medicine, syncope is called yuan xuan.
- Certain cardiac causes of syncope are connected to baroreception and output.
- In edema, the Starling equation explains the movement of fluid through cell membranes.
- In Chinese medicine, edema is considered as shui zhong.
- Claudication is a chronic condition of both legs, and it affects the ability to walk or do exercise and the duration of that ability.

INTRODUCTION

When a patient describes symptoms during the narrative, specific types may be considered a sign of cardiovascular problems. The five main symptoms include chest pain, dyspnea and orthopnea, palpitations, edema and claudication.

When a patient describes chest pain, they might state that the sensation of pressure, tightness, or discomfort radiates around the chest. There are two types of chest pain: acute and chronic. Acute chest pain is characterized by an initial onset with a short or unstable duration. The main concern is whether the episode is due to angina pectoris or myocardial infarction. Usually an EKG will show Q waves or ST segment elevations. Chronic pain is described as pain which changes in intensity frequency, and with predictable and stable episodes. The main concerns include coronary artery disease and myocardial ischemia. In Chinese medicine, when chest pain is a positive sign of pathology it may be called xiong bi or chest obstruction.

When a patient describes dyspnea and orthopnea, they may state that the chest feels uncomfortable and breathing is labored and difficult to control. Dyspnea is experienced during physical exertion and while doing regular activities. Orthopnea is a startling experience which wakes a person from sleep. Usually the breathing is difficult to control while lying down, and relief is experienced when standing upright. In Chinese medicine, both breathing problems are characterized by lung qi deficiency, with a failure of the kidney to grasp the qi. Damp phlegm obstruction causes asthma, especially when the patient has difficulty lying down.

When a patient describes heart palpitations, they may say that they are aware of strong and irregular heartbeats. Perhaps the rhythm can be tapped out on the examination table as pounding, fluttering, racing and skipping beats. Arrhythmia is categorized in Chinese medicine as xiong bi or chest obstruction, xin ji or palpitation and xuan yun or vertigo.

When a patient reports that they have experienced episodes of syncope, they will most likely need to use the observations of bystanders. Syncope is defined as a sudden loss of consciousness due to a sudden decrease in blood pressure followed by a decrease in cerebral blood flow. In Chinese medicine, syncope may be considered yuan xuan or dizziness and vertigo. Some patients may report that regularly worn clothing, shoes or jewelry are instantly tighter than normal. Some patients will notice facial puffiness.

Edema is defined as the accumulation of excess fluid in the interstitium. In Chinese medicine, edema is classified as shui zhong (shoo-ay jong). When a patient reports that they have trouble walking, they may notice a painful, cramping sensation in the muscles due to movement or exercise. If the patient attempts to continue movement, the pain will persist, but it will be relieved after 2 minutes of standing still.

Claudication can be experienced anywhere in the body; but when it occurs in the legs it is a sign of vascular disorder. Claudication affecting one or both lower legs is considered significant with other positive findings during physical examination. In Chinese medicine, claudication is considered in the category of vessel bi and blood stagnation.

CAUSES OF SYMPTOMS

1. Chest Pain

Western Medicine

Pain description (see Table 4.1).
Main causes (see Table 4.2).

TABLE 4.1

Symptom	Angina Pain Quality	Non-Angina Pain Quality
Pain	Sensation is crushing, dull or heavy	Stabbing, burning, sharp
Radiation	Involves the left, right arm or neck	Shoulders or back
Exertion	Symptoms are relieved by rest	May or may not be relieved by rest
Stress	Symptoms don't present	Emotional or sexual events will trigger

Acute chest pain (see Table 4.3).

TABLE 4.2

Pain Source	Causes	Symptoms
Cardiovascular	Ischemia	Myocardial infarction
	Non-ischemic	Pericarditis or aneurysm
Pulmonary Causes	Obstruction	Aortic disorders
	Chest wall	Pneumothorax and pleuritis
	Functional	Pneumonia, sarcoidosis, cancer, tuberculosis
	Vascular	Hypertension, embolus
	Cough	Sharp sound expelled from the lungs and throat and out of the mouth
Other Causes	Dyspnea	Difficult or labored breathing
	Psychological	Neurosis, depression, anxiety
	Gastro-esophageal disorders	Acid reflux, nausea, dysphagia and vomiting

Chronic chest pain (see Table 4.4).
Other causes (see Table 4.5).

Chinese Medicine (see Table 4.6)

2. Dyspnea and Orthopnea

Cardiovascular diseases often cycle with respiratory disorders and interfere with activity tolerance and breathing at rest. The Hering–Breuer reflex is

TABLE 4.3

Cardiovascular Cause	Factors
Unstable Angina Pectoris	The pain is relieved by rest or nitroglycerin.
	An electrocardiogram shows Q waves and ST segment depressions.
	Blood work includes CBC and cardiac enzyme test.
Myocardial Infarction	Diagnosis can include coronary artery disease.
	The pain includes dyspnea or nausea, and is not relieved by rest or nitroglycerin.
	An electrocardiogram shows Q waves and ST segment depressions.
	Blood work includes CBC and cardiac enzyme test.
Aortic Dissection	The pain includes a ripping sensation in the chest, possible absent pulse and is not relieved by rest or nitroglycerin.
	Chest X-ray is important in diagnosing.
Pericarditis	The pain radiates to the left arm and is worse while lying down and relieved by sitting up.
	An electrocardiogram shows Q waves and ST segment depressions.

then activated, causing rapid, shallow breathing, leading to an acid–base imbalance, hypoxia, anemia and oxygen deficit.

Dyspnea on exertion: the breathing is uncomfortable during regular movement, and the patient will also complain of fatigue. For example, in left ventricular failure the patient exercises and experiences dyspnea due to insufficient cardiac output and pulmonary congestion.

Paroxysmal nocturnal dyspnea: this is a condition in patients with left and right ventricular heart failure and increased pulmonary fluid pressure. The patient is suddenly awakened while sleeping in a prone or supine position. The patient obtains relief by sitting up in bed, propped up with several pillows.

Orthopnea: this condition awakens the sleeping patient who is lying prone or supine. These patients may already have failure in both ventricles. Increased pressure of fluid through pulmonary circulation decreases when the patient sits or stands up.

TABLE 4.4

Cardiovascular Cause	Factors
Coronary artery disease	The pain is relieved 3 minutes after taking nitroglycerin.
Myocardial ischemia	The pain is relieved after taking nitroglycerin, and calcium channel blockers.
	An electrocardiogram shows Q waves and ST segment depressions.
	Blood work includes CBC and cardiac enzyme test.
Mitral valve prolapse	The pain is not relieved by rest or nitroglycerin.
	Doppler and EKG is important in diagnosing, especially for vegetations in endocarditis and possible arrhythmias.
Pericarditis	The pain radiates to the left arm and is worse while lying down and relieved by sitting up.
	An electrocardiogram shows Q waves and ST segment depressions.

TABLE 4.5

Other Causes	Factors
Esophageal disorders	The pain happens as a result of movement of stomach acids: lying down or eating.
	Spasms are detected along the walls of the esophagus.
Mitral valve prolapse	The pain is not relieved by rest or nitroglycerin.
	Doppler and EKG is important in diagnosing, especially for vegetations in endocarditis and possible arrhythmias.
Pericarditis	The pain radiates to the left arm and is worse while lying down and relieved by sitting up.
	An electrocardiogram shows Q waves and ST segment depressions.

Palpitations

Palpitations are characterized as abnormal adjustments in the rate, rhythm and contractility of the heart. In most occurrences, the beat pattern includes an extra beat at systole. Without other positive signs and symptoms, it is difficult to determine whether the palpitations are due to heart disease.

TABLE 4.6

		Chinese Medicine Causes and Symptoms of Chest Pain
Stagnation	Blood	Stabbing, fixed pain, sudden onset which is worse at night
	Phlegm	Chest oppression, heavy chest, cough with mucous, fatigue and dizziness
	Cold	Pain which is worse with cold, shortness of breath, a preference for warmth
Deficiency	Qi	Fatigue, low energy and shortness of breath
	Yin	Fatigue, night sweats, afternoon fever, malar flush
	Yang	Fatigue, cold body sensation, pale face, body edema, profuse clear urination

In heart disease, factors can include septal defects, complicated mechanisms of compensation, and changes in circulation affecting mitral and aortic function.

The causes of palpitation are shown in Table 4.7.

The health practitioner must review the complete patient history and the patient narrative, as both provide useful information. Some events which trigger symptoms include unusually rapid rhythm experienced during exercise, or emotional upset. It is important to quickly determine whether there are other cardiovascular symptoms, such as chest pain or dyspnea, or existing heart diseases such as ischemia or myopathy. It is also necessary to determine the presence of coronary artery disease and hypertension. In some serious cases, syncope may be experienced before or after abrupt heart rhythm changes. When a patient explains heart rhythm changes, often they may be asked to tap the "cadence" or the beat pattern on a table surface. The quality of the rhythm provides more clues toward a diagnosis (see Tables 4.8 and 4.9).

In Chinese medicine, palpitations are characterized as xiong bi or chest obstruction, xin ji or palpitation and xuan yun or vertigo. The rhythm changes in palpitations are either tachycardia or bradycardia. Tachycardia is often differentiated as blood stagnation and qi and yin deficiency. Bradycardia is usually differentiated as blood stasis which blocks the heart vessel, phlegm, yang deficiency, cold and dampness. Health practitioners are interested in the quality of the heart beats which can be considered irregular or compensating. The detailed health history, and possibly a well organized palpitation diary of the details of events and the times of occurrences may help to determine whether the symptoms are of emotional, non-specific or specific cardiac origin.

TABLE 4.7

Palpitation Causes	Description
Non-cardiac	Some non-cardiac symptoms which occur in conjunction with palpitations include anxiety and depression. These symptoms are also part of a sedentary lifestyle. Other causes may be related to fever due to a serious illness, hypertension, and various anemias leading to cardiac output problems.
Cardiac	Palpitations are a major feature of arrhythmias, and can be experienced at rest, while sleeping, and during the process of moving from a sitting to a standing position in orthostatic hypotension.

TABLE 4.8

Palpitation Quality	Rhythm Description
Non-specific irregularities	The beats can be described as fluttering, pounding, skipping or racing.
Specific compensations	A beat described as a sinking feeling followed by unknown feelings of apathy, may correspond to an extra-systolic compensatory pause.

TABLE 4.9

Type of Arrhythmia	Quality of Rhythm Sensation
Paroxysmal atrial fibrillation	The rhythms are irregular.
Sinus tachycardia	The heart rhythm increases abruptly in a regular rapid rhythm.
Paroxysmal supraventricular tachycardia	The heart rhythm increases abruptly then terminates into regular palpitations. This is life threatening.

Syncope

In Chinese medicine, syncope may be considered yuan xuan and in the category of yun jue, which involves blood stasis with qi and blood deficiency and phlegm. The matter of diagnosing, treating and monitoring patients with well-documented cases of syncope can be a complicated process. Most patient diagnoses are often initiated in the Western medicine clinical setting with monitored referrals for evaluation and treatment with Chinese medicine.

The main physiologic factors involve impulses from baroreception, which act in vasodilation and constriction, and areas of the neurologic system such as the vagus nerve or glossopharyngeal nerve, which affect the function of the sinus node. Non-cardiac factors include vasodepressor, micturiction and defecation syncopes. An example of vasodepressor syncope is when a patient faints while having blood drawn or when experiencing emotional or psychologic situations. Micturition syncope occurs when passing urine. Defecation syncope occurs during defecation. Cardiovascular reasons for syncope include problems with output and blood flow. Syncope may be experienced in patients with arrhythmias. Ventricular failures are associated with blood flow problems. Right ventricular failure contributes to pulmonary hypertension. Insufficient cardiac output, associated with left ventricular failure, leads to decreased cerebral blood flow. Problems with the cerebral arteries – such as in ischemia or the formation of emboli – may lead to syncope. The reticular activating system (RAS) is made up of the brainstem and cortex, and it controls consciousness – being awake and alert. In pathology the RAS is affected when there is ischemia. The vertebrobasilar arteries are most affected by emboli. Resulting neurologic disorders such as stroke can lead to paralysis and death.

Edema

The Starling Equation

$$J_v = K_f([P_c - P_i] - \sigma[\pi_c - \pi_i])$$

Starling forces of hydrostatic and oncotic pressure move interstitial fluid across capillary membranes and lymphatic vessels. Oncotic pressure moves fluid out, while hydrostatic pressure moves the fluid in. Fluid accumulation, leading to edema, is caused by an increase in hydrostatic pressure; it involves, in particular, the retention of water and sodium in the kidneys, and a reduction in oncotic pressure between the tissues and within blood vessels, which causes increased permeability and obstruction within the lymphatic system, resulting in a fluid stasis.

In Chinese medicine, edema is considered as shui zhong (shoo-ay jong), is differentiated as lung wind invasion, spleen yang deficiency, spleen damp invasion, kidney yang deficiency, liver excess and heart yang deficiency. In Western medicine, this translates into the renal system, cardiovascular system and hepatic system.

In Chinese medicine, edema shui zhong (shoo-ay jong) is sometimes differentiated as spleen yang deficiency, spleen damp invasion, often differentiated as lung wind invasion, nearly always differentiated as kidney yang deficiency, liver blood stagnation and heart yang deficiency. Edema is considered a positive sign and is factored with others during physical examination. Diagnostic methods in Western medicine are needed to determine the magnitude of the condition in relationship to serious chronic diseases. There are at least three complications which are often involved in the development of localized and peripheral forms of edema: congestive heart failure, nephrotic syndrome, and liver cirrhosis. Peripheral edema indicates a chronic condition, and the cause requires careful patient analysis, diagnostic screening and close monitoring through the treatment process.

(a) Renal Factors

Starling forces will sacrifice plasma volume, with the formation of interstitial fluid accumulation, and the kidneys will retain salt and water thus contributing to nephrotic syndrome. Glomerulonephritis with hypo-albuminemia contributes to edema. Increases in arterial pressure due to kidney problems will eventually affect the lungs and heart, especially causing compensation reactions in the left ventricle. In Chinese medicine, this is differentiated as lung and kidney yang deficiency.

(b) Cardiopulmonary Factors

Cardiopulmonary disorders involved in the formation of edema include the right and left heart. Right heart failure involves the elevation of right atrial pressure which affects the lungs and lymphatic flow. Left heart failure involves the left ventricle and myocardium; systemic pressure is affected – contributing to peripheral edema. In Chinese medicine, this is differentiated as lung wind water, heart yang and qi deficiency.

(c) Hepatic Factors

Acute liver inflammation and cirrhosis create a complex systemic situation which contributes to the accumulation of ascites; this will eventually cause an extrinsic compression of the inferior vena cava. The hydraulic pressure within the capillaries will eventually contribute to edema in the extremities. In Chinese medicine, this may be seen in liver blood stagnation and kidney yang deficiency.

Claudication

Patients presenting with signs of claudication may also have atherosclerosis as a chronic condition in one or both legs. The severity of the symptoms is directly related to the degree of the stenosis and the progression rate from the age of onset. The age range of onset is between 40 and 70 years; it is more common in males than in females, and the associated risk factors are due to disease and lifestyle choices such as diabetes mellitus, hypertension, cigarette smoking and alcoholism.

Clinical Examination

Chest Pain

(a) Interview

Patient reports: The pain signals may be transmitted by the vagus and the phrenic nerves; they may also originate in and possibly be felt on palpation around the ribs and inter-costal tissues. The sensation may also be felt in the esophagus, diaphragm, scapulae or spine.

Differential diagnosis: Findings outside of cardiovascular causes may suggest pulmonary disorders or esophageal disorders such as contraction, acid reflux or esophagitis.

Genetic factors to consider (see Chapter 11, Arrhythmia):

1. DKN2A and DKN2B
2. MTAP

(b) Physical Examination

Pulses:

- Chinese pulse diagnosis (Table 4.10)

Western medicine:

- In Western medicine, the pulse may be considered 'bounding', which is similar to choppy and surging.

TABLE 4.10

	Pulses	
Stagnation	Blood	Choppy, wiry, surging
	Phlegm	Rolling
	Cold	Tight, slow, deep
Deficiency	Qi	Rapid and weak
	Yin	Rapid and weak
	Yang	Slow and weak

Auscultation:

- First palpate the apical impulse for left ventricular enlargement or hypertrophy.
- Listen for an early sign of hypertension, the fourth heart sound (S4), which indicates left atrium overwork.
- Listen for an underlying sign of heart disease, the third heart sound (S3), which indicates left ventricular malfunction.

Pathology:

- Regurgitation may be heard at the aortic position in aortic dissection.
- A pericardial friction rub may be heard in pericarditis.
- A mid-systolic click or late systolic murmur may be heard in mitral valve prolapse.

Dyspnea and Orthopnea

The patient reporting the ability to exhale, but difficulty with inhaling, most likely has a failure of the kidneys to grasp the qi. The tongue and pulse examination may or may not be a significant diagnostic method to determine diagnosis, especially when the commonly noticed signs will include a pale tongue and a weak and empty pulse. However, it is still necessary to inspect, as the pulse may change when the condition includes an exterior syndrome or a cardiovascular complication, for example.

(a) Interview

While asking the patient about symptoms, determine whether they occur or are complicated during movement, at rest or while sleeping.

Instruments used to help the practitioner ask the right questions and to evaluate breathing difficulties are listed in Table 4.11.

TABLE 4.11

Instrument	Function
Medical research scale (MRC)	Measures presence of cardiopulmonary disease.
Borg scale	Measures and evaluates effort during exercise.
Visual analog scale (VAS)	Measures breathing difficulties in degrees or levels.

TABLE 4.12

Pulses	
Lung qi	Weak
Failure of kidney to grasp qi	Deep and weak

(b) Physical Examination

Pulses:

- See Table 4.12

  Western medicine:

- Not a significant indicator

Auscultation:

- First palpate the apical impulse for left ventricular enlargement or hypertrophy.
- Listen for an early sign of hypertension, the fourth heart sound (S4), which indicates left atrium overwork.
- Listen for an underlying sign of heart disease, the third heart sound (S3), which indicates left ventricular malfunction.

Pathology:

The following cardiac sounds may be heard on auscultation along with inspecting the lungs. Pathologic sounds include regurgitation, which may suggest aortic dissection; a pericardial friction rub may be heard in pericarditis, and a mid-systolic click or late systolic murmur may be heard in mitral valve prolapse.

Palpitations

(a) Interview

Patient reports. Patients may describe the quality of the pain, any accompanying dizziness or vertigo floor rising. Chinese medicine differentiation for tachycardia or bradycardia considers the characteristics of the reported pain (see Table 4.13).

Genetic factors to consider:

1. miRNA
2. KCN familial atrial fibrillation

TABLE 4.13

Chinese Medicine Causes and Symptoms of Chest Pain		
Tachycardia	Blood stagnation	Stabbing pain
	Heart qi deficiency	Short breath, insomnia, spontaneous sweating, low voice
	Yin deficiency	Five pal heat, insomnia, dry mouth, irritability,
Bradycardia	Cold stagnation	Chest oppression, pale face, low voice,
	Blood stasis	Stabbing pain
	Phlegm fire	Restlessness, chest oppression, nausea and vomiting, scanty yellow urine, constipation

(b) Physical Examination

Blood pressure:

- This is taken in sitting, standing or supine positions, and readings indicate the stages of hypertension as a possible factor.

Pulses:

- See Table 4.14

Western medicine:

- In Western medicine, the irregular or intermittent pulse which 'skips beats' is used in diagnosing palpitations.

Auscultation:

- First palpate the apical impulse for left ventricular enlargement or hypertrophy.
- Listen for an early sign of hypertension, the fourth heart sound (S4), which indicates left atrium overwork.
- Listen for an underlying sign of heart disease, the third heart sound (S3), which indicates left ventricular malfunction.

Pathology:

- Paroxysmal atrial fibrillation and sinus tachycardia are considered in diagnostic testing. Paroxysmal atrial fibrillation is characterized by irregularities in rhythm, or skipping of beats, and missing P waves noticed on EKG. In sinus tachycardia, the sensation is a gradual onset with the heart rate at or above 100 bpm. On EKG, the P and T waves are seen close together. Both patterns may be considered a normal rhythm. When

TABLE 4.14

Pulses		
Tachycardia	Blood stagnation	Choppy
	Heart qi deficiency	Wiry
	Yin deficiency	Rapid
Bradycardia	Cold stagnation	Slow
	Blood stasis	Choppy
	Phlegm fire	Rapid and rolling

certain mild-to-moderate symptoms accompany, such as those found in hypertension and certain respiratory and endocrine disorders, then Chinese medicine can be used either alone or integrated with Western medicine therapies in treatment. Increased intensity, or 'racing' of the beat patterns through the heart, may indicate degrees of heart block and serious arterial and valvular disorders such as in supraventricular tachycardia. These are life-threatening conditions which must be monitored closely with the possibility of emergency visits to a hospital. The Valsalva maneuver can be used as a diagnostic method to detect certain heart conditions, especially in supraventricular tachycardia and aortic stenosis.

Arrhythmias:

(a) Ventricular Tachycardia (Figure 4.1)

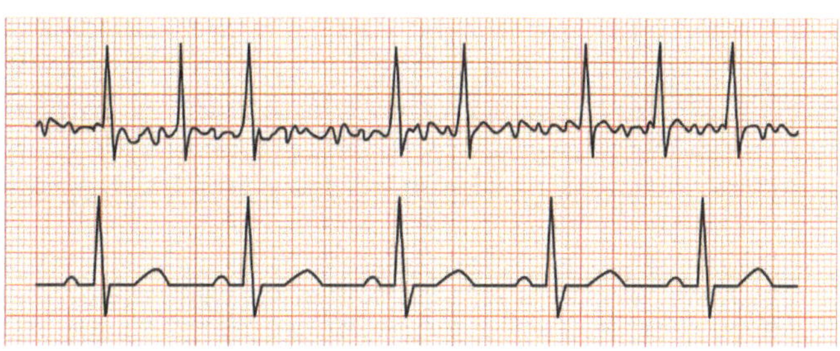

FIGURE 4.1 Ventricular tachycardia. Normal rhythm

(b) Paroxysmal Atrial Fibrillation (Figure 4.2)

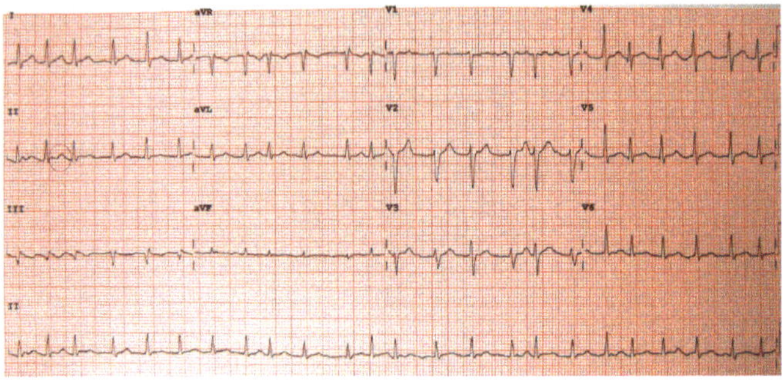

FIGURE 4.2

(c) Sinus Tachycardia (Figure 4.3)

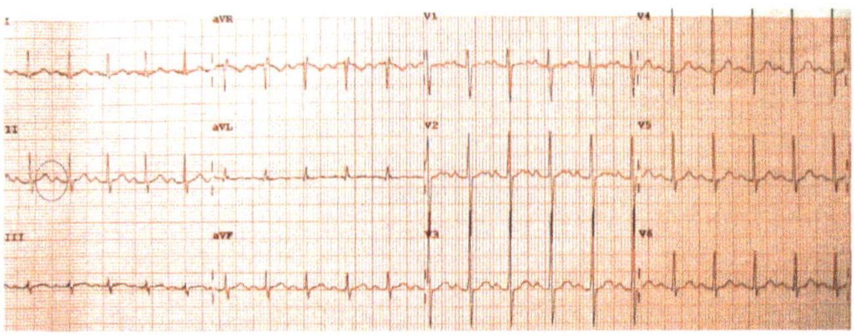

FIGURE 4.3

(d) Bundle Branch Block (atria) (Figure 4.4)

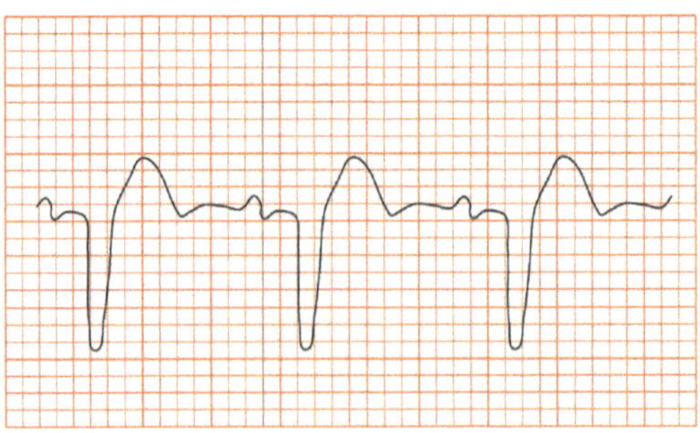

FIGURE 4.4

Syncope

The main cardiac causes of syncope include problems with output and blood flow. Arrhythmias contribute to poor cardiac output, leading to hypertension and hypotension, which are factors in episodes of syncope. Arrhythmia is characterized as either tachycardia or bradycardia, and is categorized in Chinese medicine as a deficiency state in xiong bi or chest obstruction, xin ji or palpitation and xuan yun or vertigo. In this cause of syncope, tachycardia is differentiated as blood stasis and qi and yin deficiency, and bradycardia is differentiated as blood stasis and phlegm. Syncope associated with cerebrovascular disease may be due to vertebro-basilar disorders associated with ischemia such as vertigo and sensory, visual and motor disorders. This Chinese medicine syncope may be differentiated as blood deficiency, blood stasis and hemorrhage.

(a) Interview

Patient reports: Possible reasons for underreporting symptoms involve the frequency and memory of the events. Episodes may not occur often enough for patients to consider syncope a problem. Also, witnesses may not be available to recount the occurrence. To determine whether the episodes are of cardiac or non-cardiac origin, patients may need to be referred to a Western medicine cardiac or neurologic specialist who can monitor and provide insight through additional diagnostic testing.

Non-Cardiac Causes (Table 4.15)

Genetic factors to consider:

- Genetic predispositions towards certain cardiovascular disease may be correlated with the causes of syncope in some patients.

1. miRNA
2. KCN (potassium channel) gene mutations and familial atrial fibrillation

(b) Patient Symptoms

Blood pressure:

- This is taken in sitting, standing or supine positions, and readings can indicate the presence of hypertension or hypotension as possible factors. The patient should be assisted to stand from a sitting or lying position, as fainting may occur.
- Subclavian steal syndrome. The blood pressure should be taken in both arms. A decrease in pressure >20 mmHg between the two arms may be a positive indication.

TABLE 4.15

Type of Syncope	Description
Vasodepressor syncope	This form of syncope happens in response to emotional stress concerning a real injury or a threat. It occurs when the patient rises into the standing position.
Micturition syncope	This form of syncope usually occurs after urination, and may be due to malnourishment, exhaustion, bacterial or viral infection, excessive illicit drug or alcohol intake or medications such as diuretics and orthostatic hypotension.
Defecation syncope	This form of syncope usually occurs in the elderly and may be of unknown cause.
Orthostatic hypotension	This form of syncope occurs when the blood pressure drops <90 mmHg. It may be due to central and peripheral nervous system disorders, blood volume depletion leading to decreased venous return, and certain medications.
Swallow syncope	This form of syncope is rare but can be connected with bradycardia, AV block. It occurs with esophageal diseases.
Cough syncope	This form of syncope occurs in patients with cardiomyopathies, pulmonary obstructive disease and other lung diseases who are heavy smokers and alcohol users.
Carotid sinus syncope	This form of syncope occurs in elderly patients where tight carotid sinus. Medications such as digoxin and propanolol, and also tumors, and inflammation may cause occurrence.
Exertional syncope	This form of syncope occurs in patients during brute efforts such as in moving heavy objects, playing sports and in some cases during regular movement.

Pulses:

- Common Chinese pulse diagnosis (Table 4.16)

Western medicine:

- In Western medicine, consider the cause of the syncope, especially in hypotension. Generally, the pulse may be slow and weak. However, with arrhythmias the pulse may be rapid.

Auscultation:

- First palpate the apical impulse for left ventricular enlargement or hypertrophy.

TABLE 4.16

Pulses	
Blood stagnation	Choppy
Blood deficiency	Weak
Qi deficiency	Thin
Dampness, phlegm	Rolling

TABLE 4.17

Structural Problems	
Aorta	Aortic stenosis, aortic dissection
Cardiac tissue	Hypertrophic cardiomyopathy, acute myocardial infarction
Chamber	Cardiac tamponade, and left atrial myxoma
Pulmonary	Pulmonic stenosis, pulmonary hypertension

- Listen for an early sign of hypertension, the fourth heart sound (S4), which indicates left atrium overwork.
- Listen for an underlying sign of heart disease, the third heart sound (S3), which indicates left ventricular malfunction.
- Listen for a bruit in the carotid artery, indicating a possible murmur and arterial narrowing.
- Structural problems can be assumed if a murmur is detected in auscultation. Detections using echocardiography often reveal chamber or septum dimension changes. Percutaneous intervention studies often reveal evidence of cardiac artery occlusions and other vascular changes. Other cardio-pulmonary conditions involving structural problems can include those listed in Table 4.17.

Diagnostic testing:
Episodes of syncope rarely happen at the time of clinical visits. Prolonged EKG monitoring is suggested for patients with arrhythmia and for those without a cause or history of syncope. An EEC and/or CT scan may be needed to rule out seizure from other neurologic causes of syncope. Angiography may be needed to explore or monitor carotid artery disease, and subclavian steal syndrome may require cerebral angiography.

TABLE 4.18

Symptoms of Edema		
Cardiopulmonary	Increased abdominal circumference, dyspnea, facial swelling, edema, pulmonary congestion, ascities, lower leg edema, sacral edema	Pleural effusion Lung disease Left ventricular failure Right ventricular failure
Nephritic	Fluid accumulation in the lower legs with hyperpigmentation, peripheral edema	Glomerulonephritis Congestive heart failure Nephritic syndrome
Hepatic	Ascities	Liver cirrosis
Lymphatic	Localized obstruction	Lymphadenopathy lymphangiitis

Edema

Patients with edema may need to be referred to a Western medicine specialist who can monitor and provide additional diagnostic testing and procedures for any complications.

(a) Interview

Patient reporting. The patient may present with the physical evidence of edema. The locations of the symptoms can help to determine possible causes.

Symptoms of edema:

See Table 4.18

(b) Physical Examination

See Table 4.19 for the locations of edema.

Inspect the head and face. Observe any tissue swelling located around the cheeks, around the orbital regions of the eyes, around the neck, and on sections of the tongue. The location of edema in this area of the body is differentiated as lung wind water syndrome in Chinese medicine and may be due to hereditary angioedema in Western medicine.

Palpate the bloated abdomen for ascites. While the patient is on the examination table in a supine position, stand facing their right side. Place the left hand on the left side of the waist. With the right hand pat the right side of the abdomen. Ripples in the flesh may be an indication of fluid accumulation. In Chinese medicine, this is evident in liver blood deficiency and kidney yang deficiency; in Western medicine it is evident in cirrhosis of the liver.

TABLE 4.19

Location of Edema	
Peripheral (all four extremities)	Cardiac failure renal failure
One arm	Lymphatic obstruction and lymphadenopathy, venous occlusion
Both arms	Superior vena cava occlusion
One leg	Lymphatic obstruction, deep vein thrombosis, phlebitis
Both legs	Chronic venous insufficiency, nephritic syndrome, glomerulonephritis, left and right ventricular failure

TABLE 4.20

Pulses	
Lung wind-water	Slippery and floating
Lung and kidney deficiency	Weak and deep
Heart yang (and qi) deficiency	Weak, knotted and deep
Dampness, phlegm	Rolling

Palpate the limbs for edema. Acute edema, as seen on the lower legs, will cause pitting with minimal digital pressure. Chronic edema may also present with cutaneous hyperpigmentation and dermatitis. This is physical evidence of kidney yang deficiency, nephrotic syndrome and, in some cases, right or left heart failure.

Blood pressure:

This is taken in sitting, standing or supine positions, and readings indicate stages of hypertension with edema. Fluid pressure in the vessels will raise the blood pressure. It will be necessary to notice the location of the edema to determine the possible cause.

Pulses:

- Chinese pulse diagnosis: see Table 4.20.

Western medicine:

- In Western medicine, consider the increase in pulse rate with the presence of edema.

- Kussmaul's sign: a rise in jugular venous pressure (JVP) on physical examination which indicates right heart failure. This is evident in heart yang and qi deficiency and in congestive heart failure.

Genetic factors to consider:
Genetic predisposition towards certain cardiovascular disease may be correlated with the causes of edema in some patients.

1. Hereditary Angioedema (HA)

This is an autosomal dominant condition which is due to deficiency of the C1 inhibitor protein of serpins. The symptoms of edema may be idiopathic or due to exposure to a pathogen. It is characterized by facial or tongue swelling and peripheral edema. Because of the idiopathic nature of HA in some patients it may be difficult to treat the edema, which may accumulate with associated chronic pathology of the heart, lungs, liver or kidneys. There are three types: types 1 and 2 are hereditary, and type 3 is acquired.

Claudication:

Gait is the first observance in physical examination. This section should be done especially for those with other known cardiovascular disease symptoms and for elderly patients.

(a) Interview

Patient reporting. The patient may complain of pain or difficulty walking or complying with an exercise program due to frequent cramping, which interrupts activity for several minutes at a time.

(b) Physical Examination (Table 4.21)

Gait. The trip from the waiting room to the examination should provide evidence of difficulty. Also it should provide enough travel distance or amount of time before the occurrence of symptoms.

TABLE 4.21

Location of Claudication	Occluded Artery
Greater trochanter	Iliac artery and profunda femoris
Thigh	Ilio femoral artery segment
Calf	Femoral artery
Foot	Tibial arteritides

Mounting or dismounting the examination table: The muscle groups of the legs work together to help the patient on or off the table. If there is pain in movement, the patient will need assistance.

Blood pressure and heart rate:

This is taken in sitting, standing or supine positions, and readings indicate stages of hypertension which are also associated with claudication in some patients. An increase in pulse rate is also apparent in the presence of claudication.

2. Genetic Factors to Consider:

Genetic predisposition towards certain cardiovascular diseases may be correlated with the predisposition of claudication in some patients.

Parental Intermittent Claudication The children of parents with intermittent claudication may likely develop claudication and peripheral artery disease during middle age.

SUMMARY IN THE CONTEXT OF PERSONALIZING MEDICINE

If the causes of the symptoms are cardiac it may be helpful to consider the severity, and any possible genetic predispositions. Also, it will be necessary to interview the patient, perform physical examinations, and get results of certain diagnostic tests. All of the information will help determine whether the patient is indicated for treatment using Chinese medicine alone, integrative medicine treatments or Western medicine alone.

FURTHER READING

[1] Goodacre S, Cross E, Arnold J, et al. The health care burden of acute chest pain. Heart 2005;91:229–30.

[2] Goodacre S, Nicholl J, Dixon S, et al. Randomised controlled trial and economic evaluation of a chest pain observation unit compared with routine care. BMJ 2004;328:254–7.

[3] The MRC breathlessness scale. Occup Med (Lond) 2008;58(3):226–7. Available from: http://dx.doi.org/doi:10.1093/occmed/kqm162.

[4] Clark AL, Poole-Wilson PA, Coats AJ. Exercise limitation in chronic heart failure: central role of the periphery. J Am Coll Cardiol 1996;28(5):1092–102.

[5] Lim SH, Anantharaman V, Teo WS, et al. Comparison of treatment of supraventricular tachycardia by Valsalva maneuver and carotid sinus massage. Ann Emerg Med 1998; 31(1):30–5.

[6] Freeman R, Wieling W, Axelrod FB, et al. Consensus statement on the definition of orthostatic hypotension, neurally mediated syncope and the postural tachycardia syndrome. Auton Neurosci 2011;161:46–8.

[7] Shah MG, Cho S, Atwood JE, Heidenreich PA. Peripheral edema due to heart disease: diagnosis and outcome. Clin Cardiol Jan 2006;29(1):31–5.

[8] Smith GD, Shipley MJ, Rose G. Intermittent claudication, heart disease risk factors, and mortality. The Whitehall Study. Circulation, Dec 1990;82(6):1925–31.

Chapter 5

Hematology

Chapter Objectives

After studying this chapter you should be able to:

1. Explain the Chinese medicine and Western medicine perspectives of blood
2. Explain the in-flow and circulation of qi from the Chinese medicine and Western medicine perspectives
3. Discuss the stages and levels of immunity and pathology in Chinese medicine, and imagine the connection between it and the role of the lymphatic system and white blood cells
4. Discuss the laboratory tests used and the significance of using them in an integrative medical setting in cardiovascular disease management

OVERVIEW

This chapter describes hematology through the perspective of Chinese medicine and Western medicine. The integrating theories and sciences section includes the two types of essence, the zang organs involved in blood production and movement in Chinese medicine, the components which make up blood, the types of qi, and the significance of the Bohr effect in qi and genetic blood problems which predispose towards cardiovascular disease. The immune system section includes the six stages and four levels in Chinese medicine to explain the penetration of a pathogen through the layers of the immune system and the levels of pathology that it will cause, the role of the lymphatic system, and the development of leukocytes and lymphocytes. The laboratory test section includes specific diagnostic tests, the inclusion of which may be important for the cardiovascular patient.

Key points

- Innate and acquired are two types of essence in Chinese medicine.
- The production and movement of blood is contingent upon the efforts of zang organs.

Integrative Cardiovascular Chinese Medicine. DOI: http://dx.doi.org/10.1016/B978-0-12-420014-2.00033-8

- Erythrocytes have eight stages of development.
- Hemoglobin makes up about 95% of blood and transports iron, carbon dioxide and oxygen.
- Plasma makes up 55% of blood volume, and is the fluid which contains all blood components.
- Thrombocytes circulate with blood and are useful for the repair of injured blood vessels.
- Genetic blood problems include thalassemia and hemochromatosis.
- Qi represents the basic life energy in Chinese medicine.
- The Bohr effect in Western medicine can be used as an example of the role of pectoral qi.
- The immune system can include aspects of the six stages and four levels in Chinese medicine, the lymphatic system, leukocytes and lymphocytes.

INTEGRATING THEORIES AND SCIENCES

Essence (see Table 5.1)

Blood

In Chinese medicine, blood is a yin substance. Proper generation and circulation is contingent upon the smoothness of the vessels, temperature, and cooperation between the spleen, liver, lung and heart zang (see Table 5.2).

Erythrocytes are red blood cells which make up about 50% of blood volume. Three million cells are produced continually each second from stem cells located in bone marrow. Once mature, erythrocytes circulate for about 100–120 days. There are eight stages of development (see Table 5.3).

Hemoglobin is the protein in blood which is rich in iron and gives it the red color. It occupies about 95% of the surface area of a red blood cell and has an oxygen-binding capacity of 1.34 mL O_2/g, which satisfies the need

TABLE 5.1

	Essence in TCM
Innate essence	This type of essence is inheritied from parents after conception. The lifetime supply is stored in the kidneys after birth to assist with body development.
Acquired essence	This type of essence is derived from regular nutritional intake of food and water.

TABLE 5.2

Spleen zang	Spleen qi controls blood formation and movement through the vessels.
Liver zang	The liver stores and controls the volume and free movement of blood.
Lung zang	Lung qi mixes fresh air with returning blood to form pectoral qi which is then moved through the heart.
Heart zang	Heart qi propels the blood through the vessels out to the rest of the body.

TABLE 5.3

	Erythrocyte Development
1. Hemo cytoblast	The early stage of a nucleitic stem cell, which could later become any kind of blood cell.
2. Myeloid	A precursor to a kind of red blood cell.
3. Unipotent stem cell	A cell which is designated to be a certain type of blood cell.
4. Pronormo blast	The precursor stage of an erythrocyte.
5. Erythroblast	The early stage of an immature erythrocyte.
6. Polychromatophilic normoblast	An immature erythrocyte with hemoglobin.
7. Ortho chromatic normoblast	The early stage of a reticulocyte, which enters the blood stream.
8. Reticulcyte	An immature circulating red blood cell.

for the average adult consumption of 250 mL of oxygen per minute. Hemoglobin transports oxygen (O_2), carbon dioxide (CO_2), nitric oxide (NO) and hydrogen (H^+). The heme groups carry the iron in hemoglobin. A heme group contains one iron atom which bonds to the nitrogen molecule of a histidine residue on one end, and on another end it is bound to one oxygen molecule, allowing a hemoglobin protein to bind to four oxygen molecules.

Plasma is the yellowish clear fluid component which includes protein, antibodies, lipids, vitamins, minerals, clotting factors, hormones, sugars, enzymes and saline solution.

Thrombocytes are also called 'platelets'. They circulate within the blood with a lifespan of about 10 days. They are useful in cases of injury, and will

TABLE 5.4

Thalassemia Type	Hemoglobin Type	Affected Decent Populations
Alpha thalassemia major	Alpha globulin	Middle eastern
		Chinese
		South east Asian
		African
Beta thalassemia minor	Beta globulin	Mediterranean

adhere at the site of a wound or to the walls of damaged vessels to release chemicals which clot and close the wound or vessel. After injury to a vessel, the thirteen different clotting factors are immediately triggered in a cascading sequence.

Genetic Problems in Blood

(a) Thalassemia

Thalassemia is a genetic blood disorder in which the body makes an abnormal form of hemoglobin that destroys erythrocytes, causing anemia. There are two main types of thalassemia. The major type leads to the development of liver and heart failure (see Table 5.4).

(b) Hereditary Hemochromatosis

Hereditary hemochromatosis involves a mutation in the *C282Y* and *H63D* genes; it primarily affects males. It involves an excessive accumulation of iron, which causes damage to the heart, liver, joints and testes.

Qi

Qi represents the basic energy which animates regular life processes and is created partially from acquired essence. It moistens the skin surface to allow anti-pathogenic qi to remain on active duty, and helps to maintain body temperature, internal blood circulation and the secretion of body fluids. There are four main types of qi: primordial, pectoral, nutritive and defensive. Two additional types are controlling qi and propelling qi.

The Bohr effect is the gas-exchange mechanism in the body. Oxygen, which is needed by all cells and tissues in the body, remains bound to hemoglobin in the arteries but it is released in capillaries. Here, the release of O_2

causes the binding of CO_2 to hemoglobin. In the lungs, the release of CO_2 causes the binding of O_2 to hemoglobin.

THE IMMUNE SYSTEM

The purpose of the six stages (Table 5.5) and the four levels (Table 5.6) in cardiovascular Chinese medicine is to consider the patterns and levels in the body, which are associated with pathogenic invasion in such patients. For example, pathogenic invasion and the resulting conditions – such as in pneumonia – can have a serious effect on the lungs. Complications have the potential to reduce the amount of available oxygen which may lead to cardiac and systemic complications. Inflammation is part of the innate immune response to certain pathogens.

The functions of the lymphatic system are to assist the immune system by stimulating an immune response, removal of interstitial fluid from tissues, and the transportation of white blood cells. The anatomy of the lymph node includes connective tissue and afferent and efferent compartments called lymph nodules. The afferent vessel allows lymph fluid to enter and flow through the node. The fluid exits the lymph node through the efferent lymphatic vessel, located at the hilum. The walls of the lymph capillaries are composed of endothelium, which consists of overlapping squamous cells; the endothelium prevents lymph fluid leaking from the vessel. The lymphatic system lacks a pumping mechanism, such as that found in the cardiovascular system, so that the movement of lymph depends on respiratory movement, skeletal muscle movement.

Leukocytes are white blood cells which make up 0.5–1% of blood, liver, spleen, and lymph nodes. They are produced in the bone marrow in stem cells, and in the thymus gland, and they have a lifespan of about 20–36 hours. Leukocytes are part of the defense or immune system. They are also called white blood cells, which signal the presence of a pathogen. The reason for studying leukocytes in cardiovascular Chinese medicine is to identify the conditions in the body for inflammation and pathogenic invasion during the six stages and four levels; this information can be used in a strategy for treatment using Chinese medicine or integrated Chinese and Western medicine. Leukocytes defend the body against invasion by pathogens. An increase in the number of leukocytes is called *leukocytosis*, and a decrease is called *leucopenia*. Leukocytes are classified as either granulocytes or agranulocytes (see Tables 5.7 and 5.8).

Lymphocytes are white blood cells which make up about 0.5–1% of circulating blood. They are produced in the stem cells in bone marrow. Mature circulating cells produce antibodies to antigens on certain bacteria, viruses and parasites, which help establish immunity to future invasion by the same kind of pathogenic organism.

TABLE 5.5

The Six Stages		
Stage	**Channels**	**Symptoms**
Tai Yang stage	– Bladder – Small intestine	Fever, chills, aversion to cold, headache
Treatment principle: Release the exterior		Tongue = thin white coat Pulse = floating, tight
Shao Yang stage	– San Jiao – Gallbladder	Alternating hot and cold sensations, headaches, dizziness, blurred vision, hypochondriac pain, irritability, bitter mouth taste, nausea
Treatment principle: Soothe hyperactivity		Tongue = thin yellow and white coat Pulse = wiry
Yang Ming stage	– Stomach – Large intestine	The 5 bigs: fever, sweat, irritability, thirst and pulse
		Others: aversion to heat and restlessness
Treatment principle: Clearing heat		Tongue = dry yellow coat. Pulse = rapid
Tai Yin	– Lung – Spleen	Abdominal pain, vomiting, loss of appetite, diarrhea
Treatment principle: Warm the cold		Tongue = dry yellow coat. Pulse = wiry
Shao Yin	– Heart – Kidneys	Hot above (yang) = fever, insomnia, dry mouth
		Tongue = red w/thin white coat Pulse = rapid
Treatment principle: Warm the cold		Cold below (yin) = cold legs, body aches, frequent clear urination and diarrhea
		Tongue = pale w/white coat Pulse = weak and rapid
Jue Yin (Shock, separation of yin and yang)	– Liver – Pericardium	Hunger and unable to eat due to vomiting, alternating sensation of hot and cold, body weakness, possible parasites
Treatment principle: balance yin and yang		Tongue = white coat with papillae Pulse = deep and hidden

TABLE 5.6

The Four Levels		
Level	**Syndrome types**	**Symptoms**
Wei level Tai yang stage	Wind heat Dry heat Summer heat Damp heat	Aversion to wind, fever, chills, headache, sweating, thirst, sore throat, dry mouth w/thirst
Exterior heat		Tongue = thin white coat Pulse = floating, tight
Qi level Yang ming stage	Lung and stomach heat Intestinal heat Spleen damp heat	The 5 bigs: fever, sweat, irritability, thirst and pulse Others: aversion to heat and restlessness
Interior heat		Tongue = dry yellow coat Pulse = rapid
Ying level Nutritional stage	Heat attack of the heart Heat attacking the pericardium	Irritability, mental/emotional disturbances, night fever, insomnia, dry mouth and thirst
Heat and fire		Tongue = red Pulse = rapid
Xue level	Liver and chong heat Severe heat and wind Severe heart heat Severe blood heat	Irritability, alternating emotional states, dry skin, rashes or twitching, epistaxis, various bleeding disorders
Attack on blood		Tongue = red Pulse = rapid, wiry

TYPES OF WHITE BLOOD CELL

Laboratory Tests

Complete Blood Count (CBC)

The purpose of the CBC is to determine the cause of complaints about fatigue or weakness, to determine the amount of blood loss during a hemorrhage, to determine whether there is an infection or inflammation present, to view the affects of certain medications or herbal formulas on the blood or immune system, or to diagnose polycythemia or anemia (see Table 5.9).

TABLE 5.7

Leukocyte	Type	Description
Neutrophil	Granulocyte	– Defense against bacterial or fungal infection. – Most common in the early acute inflammation. – Makes up 60% of total leuko cyte count. – Lifespan is 5 days.
Eosinophil	Granulocyte	– Inflammatory cells. – Responds in allergic reactions and parasitic invasions.
Basophil	Granulocyte	– Responsible for allergic reactions. – Releases histamine which causes vasodilation.
Monocyte	Agranulocytes	– Known as macrophages responsible for phagocytosis of pathogenic cells. – Shows pathogens to T cells which kill them.

TABLE 5.8

Leukocyte Disorders			
	Leukocytosis		Leukopenia
Neutrophilia	– Increase in count – Caused by primary disease types	Neutropenia	– Decrease in count. – Often drug induced.
Eosinophilia	– Presence in circulation is not normal. – Incidence may happen without cause. – May be seen in the blood with allergies, morbid cholesterol levels, and interstitial nephropathy.	Lymphocytopenia	– Decrease in count. – Incidence in immune compromised patients. – Often drug induced. – May be due to zinc deficiencies

Cardiac Markers

Cardiac biomarkers are used to detect emergency acute cardiac events and to monitor while under hospital observation (see Table 5.10).

Lipid Panel

The lipid panel provides information on TC (total cholesterol), LDL (low-density lipoprotein), HDL (high-density lipoprotein) and triglycerides. These

TABLE 5.9

Test	Reference Interval	Interpretation
WBC Per cubic mL	5000–10000	Indicates the presence of infection or inflammation.
RBC Millions/mcL	Men: 4.4–5.5 Women: 4.0–5.9	– Low count indicates a hemorrhage, anemia or iron deficiency. – High count indicates clotting as in atherosclerosis.
Hemoglobin (HgB) g/L	Men: 140–174 Women: 120–160	– Measures the ability of RBC to carry oxygen. – A good test for anemia.
Hematocrit (HCT)	Men: 42–52% Women: .36–.48%	– Measures the percent of erythrocytes in the blood. – A good test for anemia.
Mean corpuscular volume (MCV) cubic mm	82–98	– Size of the red blood cells.
Mean corpuscular hemoglobin (MCH) cubic mm	26–34	– Amount of haemoglobin in RBCs.
Red cell distribution width (RDW)	11.5–14.5%	– Compares the size and shapes of all RBCs.
Platelets		– Low count indicates idiopathic thrombo cytopenic purpura and splenomegaly.
Mean platelet volume (MPV)	Range: 9.7–12.8/ femtoliter (fL) Normal: 7–11	– Measures the average size of the platelets.
Neutrophils	50%	These tests are used to measure the presence of specific infections and inflammation.
Basophils	0–1%	
Eosinophils	0–3%	
Neutrophils	3–6%	
Lymphocytes	25–40%	
Monocytes	3–7%	
Reticulocytes	Measures the count of new red blood cells, which may indicate that mature ones are depleting fast due to destruction or a hemorrhage.	
Erythrocyte sedimentation rate (ESR)	Measures the rate erythrocytes settle inside of the test tube. The cells settle slowly with the presence of inflammation.	

TABLE 5.10

Cardiac Biomarkers		
Marker	**Description**	**Indication Period**
Troponin I and T	– Acute myocardial infarction – Heart failure – Myocarditis	– Elevates in 2–4 hours – Peaks in 12 hours – Must be measured within 3 days – May persist for 7 days
Creatinine Kinase (CK-MB)	– Diagnosis of acute myocardial infarction	– Releases immediately – Peaks within 24 hours – Normal in 2 days
Pro-brain naturietic Peptide	– Heart failure indication – Poor prognosis value = >100 – Better prognosis value = <80	Timeframe is ongoing
Glycogen phosphorylase isoenzyme BB	– Detects ischemia and MI – Early indicator in acute coronary syndrome – Detects unstable angina	– Elevates in 1–3 hours – Persists up to 7 hours

tests help to determine the levels of certain types of lipid. Sometimes the glucose test is included; this helps to monitor the blood sugar levels. Both types of test help to predict, diagnose and monitor cardiovascular disease (see Table 5.11).

Inflammation Tests

(a) Fibrinogen

- High levels of fibrinogen are markers for inflammation; this indicates an increased risk of cardiovascular disease or an impending event, and glomerulonephritis.

(b) Homocysteine

Homocysteine is an amino acid formed during methionine metabolism. High levels of homocysteine are involved in cardiovascular diseases.

(c) C-reactive protein

C-reactive protein is a marker of systemic inflammation; it is a risk factor for cardiovascular diseases such as hypertension, coronary artery disease, diabetes mellitus, and metabolic syndrome. (See Table 5.12.)

TABLE 5.11

Lipid Panel	
Glucose	Range: 65–99 mg/dL
	Normal: 70–85 mg/dL
Cholesterol	Range: 100–199 mg/dL
	Normal: 180–200 mg/dL
LDL	Range: 0–99 mg/dL
	Normal: Under 100 mg/dL
HDL	Range: 40–59 mg/dL
	Normal: Over 55 mg/dL
Triglycerides	Range: 0–149 mg/dL
	Normal: Under 100 mg/dL
Hemoglobin	Range: 4.5–5.7%
	Normal: 4.5%

TABLE 5.12

Inflammation Test	
Fibrinogen	Range: 193–423 mg/dL
	Normal: 295–369 mg/dL
Homocysteine	Men
	Range: 4.3–15.3 μmol/L
	Normal: < 7.2 μmol/L
	Women
	Range: 3.3–11.6 μmol/L
	Normal: < 7.2 μmol/L
C-reactive protein	Men
	Range: 0–3 mg/L
	Normal: < 0.55 mg/L
	Women
	Range: 0–3 mg/L
	Normal: <1.5 mg/L

TABLE 5.13

Hypothyroidism	Hyperthyroidism
• hypercholesterolemia	• atrial fibrillation
• weakness and fatigue	• heart palpitations
• low grade depression	• excessive weight loss
• cold intolerance	• nervousness
• weight gain	• insomnia
• dry skin	• heat intolerance
• constipation	• weakness and fatigue

Hormone Tests

(a) Dehydroepiandrosterone (DHEA)

DHEA is produced by the adrenal glands, and is a precursor of male and female reproductive hormones. Its presence, at balanced levels, assists in the prevention of cardiovascular diseases.

(b) Thyroid-Stimulating Hormone (TSH)

TSH is produced in the thyroid gland. Symptoms of pathology may occur as a result of hypothyroidism (low levels of TSH) or hyperthyroidism (high levels of TSH). (See Table 5.13.)

(c) Estrogen

Estrogen is a class of hormone which includes estriol, estradiol and estrone. It is a female hormone which is present in both males and females, but it is present in much higher levels in women. Balanced estrogen levels may help to decrease the risk of certain cardiovascular diseases, such as atherosclerosis.

(d) Testosterone (free)

Testosterone is produced in the adrenal glands in males and females. It is produced in higher levels within the testes of males, and in lower levels within the ovaries in women. In men, low levels of testosterone are associated with emotional problems, obesity and cardiovascular disease. In women, low levels of testosterone cause fatigue and decreased mood and libido.

SUMMARY IN THE CONTEXT OF PERSONALIZING MEDICINE

The blending of basic theories, sciences and general laboratory findings in Chinese medicine and Western medicine, are tools for health practitioners of different medical systems to understand each other. The information can be used as a guide for indicating treatments using Chinese medicine alone, integrative medicine treatments, or Western medicine alone.

FURTHER READING

[1] <http://www.webmd.com/heart-disease/cardiac-enzyme-studies>.

[2] Lubin BH. Reference values in infancy and childhood. In: Nathan DG, Oski FA, editors. Hematology of Infancy and Childhood. 2nd edition. Philadelphia, Pa, USA: WB Saunders; 1981. p. 1552–74.

[3] Kazazian Jr HH. The thalassemia syndromes: molecular basis and prenatal diagnosis in 1990. Semin Hematol 1990;27(3):209–28.

[4] Rodeghiero F, Stasi R, Gernsheimer T, et al. Standardization of terminology, definitions and outcome criteria in immune thrombocytopenic purpura of adults and children: report from an international working group. Blood 2009;113(11):2386–93.

[5] Zheng G, Schaefer M, Karplus M. Hemoglobin Bohr effects: atomic origin of the histidine residue contributions. Biochemistry 2013;52(47):8539–55.

Chapter 6

The Systems of Chinese Medicine and Western Medicine

Chapter Objectives

After studying this chapter you should be able to:

1. Explain qi activity within the cells of the myocardium and conduction system
2. Describe the function of qi within the blood vessels
3. Discuss the Chinese medicine perspective on heart pathology
4. Explain the anatomic and physiologic connections between the heart and the zang vicera
5. Describe the three main symptoms of cardiovascular disease

OVERVIEW

This chapter outlines the foundation of integrative cardiovascular Chinese medicine. Both Chinese medicine and Western medicine view the heart and cardiovascular system similarly, but from two different aspects. In Chinese medicine, shi zheng (shuh-jung) is the basis of combining the characteristics of diseases with syndrome differentiation. It is important in diagnosis, treatment and management. Blending the basic medical sciences, theories and factors from the medical science level to clinical evaluation provides the direction for providing personalized medicine.

The heart is considered to be the monarch of all zang and fu organs in the Chinese medicine system. Oxygen-depleted systemic blood, which had previously provided oxygen and nutrients to all cells, tissues and organs throughout the body, enters the cardiopulmonary section. Within this section, the blood is re-oxygenated and is then circulated back through the body in a continuous process.

The cardiac conduction system demonstrates the dynamics of qi, yin and yang within the mechanisms which motivate automaticity within ion channels,

Integrative Cardiovascular Chinese Medicine. DOI: http://dx.doi.org/10.1016/B978-0-12-420014-2.00004-1

resting potentials, action potentials and pacemaker actions. Ions are represented as yin and yang types. Their organized and synchronized movements of diffusion across a cell membrane generate the electrical energy which is heart qi. Ion channels include voltage-gated channels and receptor-gated channels of the sinoatrial and atrioventricular nodes. These channels regulate the movement of ions in and out of cells. Resting potentials determine the electrical activity of myocytes at rest before engagement. Action potentials involve the sudden depolarization episode during the ejection of blood from the atria to the ventricles. Pacemaker actions involve baroreceptors in the blood vessels and the sinoatrial node, which responds to any increase in pressure.

Blood-vessel dilation and constriction are dominated by heart qi, which also propels the blood through the vessels to deliver oxygen and nutrients first to the heart and then to the rest of the body. The vessels which serve the heart include the coronary arteries. The aorta is the vessel which sends oxygenated blood back into regular circulation.

Factors in heart pathology include ischemia and inflammation. Ischemia includes xiong bi (shee-ong bee) or chest bi in Chinese medicine. Obstructed blood flow in the coronary arteries reduces blood supply and causes conduction problems. Inflammation is an acute or chronic condition, caused by the six pathogens of Chinese medicine and syndromes of dampness and phlegm, which include vascular changes caused by inflammatory mediators with the immune system.

Pathologic factors between heart and zang/fu organs include the direct relationship through the connecting arteries. Organs are affected according to various pressures within the heart, and to the sufficiency of blood and oxygen.

Key symptoms of cardiovascular disease include chest pain, dyspnea and palpitations. These symptoms determine the indication and contraindications involved when considering integrating Chinese medicine and Western medicine.

Key Points

- The heart muscle tissue cells are composed of myocytes with abundant mitochondria which use oxygen to generate energy.
- The right atrium also houses the first part of the conduction system, the sinoatrial node (SAN), which is located in the upper section near the superior vena cava.
- Heart qi state represents the magnitude and pathology of the cell membrane potentials and conduction action potentials.
- Ca^{2+}, Na^{+} and K^{+} can be considered yang ions even though they may have interchangeable yin and yang roles during the resting potential process.
- Blood is considered yin. Blood includes both yin and yang components. Yin components include water and other fluids, and yang components include oxygen, carbon dioxide, hormones, nutrients, biochemicals and microbes.

- The dilation and constriction of the blood vessels are also dominated by heart qi.
- In Chinese medicine, ischemia is considered to be a chest obstruction or chest bi. Blood flow is decreased or blocked within the coronary arteries, reducing oxygen supply to the muscle tissues and causing problems with the conduction system.
- In Chinese medicine, acute and chronic inflammatory responses are realized through the theory of pathomechanism and syndrome differentiation.
- The immune responses to acute inflammation involve leukocytes, plasma-derived mediators and complement systems, which can be associated with the Chinese medicine six pathogenic factors with six stages experienced on four levels.
- In both Chinese and Western medicine, chronic inflammation is the long-term result of persistent infection or immune response after acute inflammation, and may be linked to hyperlipidemia and obesity.
- The superior mesenteric artery is one of the three major arteries which branch from the abdominal aorta, and it connects the small intestine zang to the heart zang.
- The splenic artery is the vessel which connects the spleen zang to the heart zang.
- The lung zang and heart zang are connected through the return of oxygen-poor systemic blood to the lungs through the pulmonary artery, and the return of oxygen-rich blood to the heart through the pulmonary veins.
- The kidney zang and heart zang are connected through the renal artery.
- The liver zang and heart zang are connected through the hepatic artery, which also receives 75% of blood flow from the spleen, pancreas and stomach.
- The brain sui is connected to the heart zang through four interconnecting arteries.
- In Chinese medicine, chest pain is called xiong (she-yong) tong. The main concerns of xiong bi are whether the pain is involved with angina or a myocardial infarction.
- In Chinese medicine, dyspnea is called qi chuan (choo-an), which is characteristic of uncomfortable and labored breathing. There are two factors involved: cardiac and pulmonary.
- In Chinese medicine, palpitations are called xin ji (sheen-jee), and are characterized by the awareness of the rate and rhythm of heartbeats.

THE HEART

See Figure 6.1.

In Chinese medicine, heart qi is a general term for the abundant yang activity expressed through electrical energy generation and movement within the heart. Heart qi is involved in mechanisms such as the in-action potentials, contractility, cardiac output, stroke volume, etc. which affect the state of the cardiovascular system as well as other organ systems.

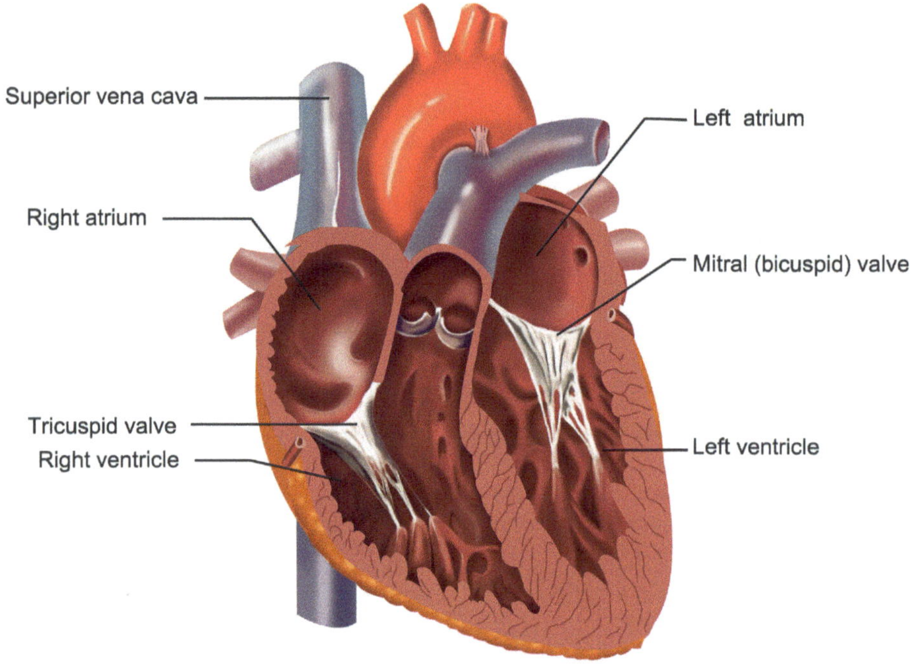

FIGURE 6.1

The muscle tissue of the heart is composed of cells called myocytes which generate qi through a weak electrical charge. The myocytes pulsate in unison, giving mild contraction effects. This contraction effect is useful as a final life-saving effort should the conduction system fail. Like skeletal muscle cells, these cells also have abundant mitochondria which use oxygen to generate energy in the form of adenosine triphosphate (ATP). This energy synthesizes macromolecules used for pumping Na^+ and K^+ ions. The ion-pump is responsible for nerve and muscle functioning, and failure leads to hypoxia. During circulation, the metabolism of oxygen results in CO_2 and the production of H^+ ions – both of which are acidic byproducts in constant need of removal. Circulating venous blood enters the heart through the superior and inferior vena cavae. It then flows through sections of the cardiopulmonary system for waste removal, replenishment of oxygen and then back through the heart; it goes out into regular circulation to nourish organ systems and tissues throughout the body.

Right and Left Atria

The right atrium is the first chamber of the heart to receive deoxygenated and carbon dioxide-rich systemic blood from the body, and the left atrium is the first chamber to receive the oxygenated blood returning from the lungs. The right atrium also houses the first part of the conduction system, the sinoatrial node (SAN), which is located in the upper section near the superior vena cava. The SAN is made up of pacemaker cells which polarize to generate an action potential. During diastole, the relaxed right and left atria fill with blood. Once filled to capacity during systole, the SAN causes them to contract simultaneously, thus pushing the blood through the tricuspid and mitral valves into their respective right and left ventricles.

Right and Left Ventricles

These chambers receive blood during systole. The right ventricle is located inferior to the right atrium and receives deoxygenated and carbon dioxide-rich systemic blood. The left ventricle is a chamber located inferior to the left atrium and receives oxygenated blood. During diastole the ventricles relax to release blood out of the heart. During systole, the right ventricle contracts and pushes blood through the pulmonary valve to the lungs to release the carbon dioxide and include fresh oxygen. The left ventricle contracts and the aortic valve opens to push oxygenated blood through the aortic valve, through the aorta, to provide oxygenated blood to the body cells and sections of the heart.

THE CARDIAC CONDUCTION SYSTEM

See Figure 6.2.

In both Chinese medicine and Western medicine, the purpose of the conduction system is to provide the energy and electrical activity for muscle cell contraction to maintain the continuous pumping action of the heart. This functioning involves heart qi, which is the product of the yin and yang activity of ions generating electrical currents within myocardial cells, and involved in contractility. The heart qi is represented in the cell membrane potentials and conduction action potentials. Yin and yang balance is represented by the characteristics and movements of ion channels, and is expressed through the phases of the resting and action potentials.

Ion Channels

In cardiac myocytes, ions enter and exit through specialized gates or channels called voltage-gated channels and receptor-gated channels. Some potassium, sodium and calcium gates are voltage-gated channels.

1 Voltage-Gated Channels

Through several channels – such as calcium-activated channels, ATP-sensitive channels, delayed, inward and inward rectifiers and outward

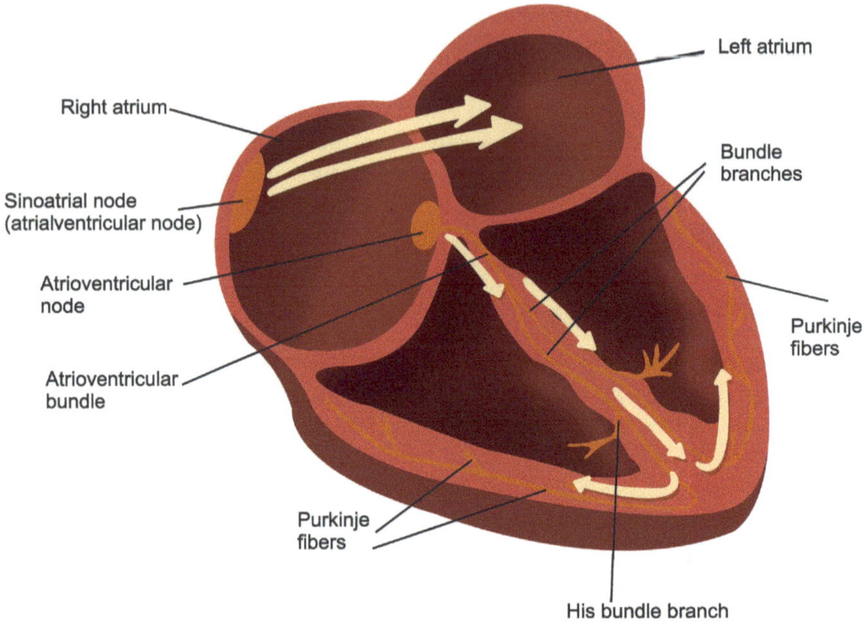

FIGURE 6.2

transient channels – K^+ can maintain a negative electrical potential as the voltage gate closes with depolarization in phase 4, contributes to activity in phase 1, and contributes to repolarization in phase 3 of the action potentials and slow sinoatrial (SA) node firing during certain conditions. When slow- or fast-gated sodium channels are open, Na^+ can contribute to the electrical current in the cells of SA and atrioventricular node (AV) nodes and to myocyte rest in phase 0. L-type or T-type Ca^{2+} channels open to provide calcium ions for long-lasting electrical current within the cell in phases 0 and 4 of the cells of the SA and AV nodes, and in phase 2 of cardiac myocytes.

2 *Receptor-Gated Channels*

These channels open and close in response to certain chemical receptors. Acetylcholine binds to cell receptors, helping to open potassium channels. Acetylcholine is also a neurotransmitter which is released by the vagus nerve. The vagus nerve is cranial nerve 10, which is a parasympathetic innervation of the SA node. It helps to lower the heart rate.

Cell Membrane Resting Potentials

Manifestation of heart qi, or phases of the generation of small-voltage electrical potential, is an important phase in the conduction system. It can determine the electrical activity of myocytes at rest, before engagement, and the balanced coordination of contraction. Resting potentials can be described by two mechanisms: the equilibrium potentials of ion gradients and Na^+/Ca^{2+} exchange. Equilibrium potentials compare the concentration gradients of negative ions (Cl^-) maintained within the cell and positive ions (Ca^{2+}, Na^+ and K^+) diffusing across (inside and outside) the cell membrane. Here, Ca^{2+}, Na^+ and K^+ can be considered yang ions, even though they may have interchangeable yin and yang roles during the resting potential process. The main yang ion gradient is K^+ because cardiac cell membranes are more permeable to it, and also the concentration of this ion inside or outside the cell determines the resting and action potentials. Within myocardial cells, the electrogenic Na^+/K^+-ATPase pump system maintains the balance of three Na^+ for every two K^+ entering the cell membrane. During the action potentials, Ca^{2+} is the important yang ion within the cell. Na^+/Ca^{2+} exchange is involved in contraction and relaxing. Three Na^+ ions are exchanged for every one Ca^{2+}, which can also influence Na^+ concentration. For example, when a drug such as digoxin increases both Na^+ and Ca^{2+} it leads to force increased contraction of cardiac myocytes.

Cell Membrane Action Potentials

Action of the heart qi in the conduction system includes pacemaker and non-pacemaker actions and happens with the sudden depolarization and repolarization back to the resting phase.

1. Non-pacemaker action potentials. These potentials represent the accumulation of and future movement of qi energy which depends on the functioning between adjacent cells along the Purkinje fibers and within the myocytes of the ventricular and atrial chambers. Phase 4 is the resting phase, and 0, 1, 2 and part of phase 3 are collectively the long effective refractory period (ERP). During this period, depolarization gives both atria time to fill with returning blood and to prepare for repolarization with ejection into the ventricles.

   See Figure 6.3.

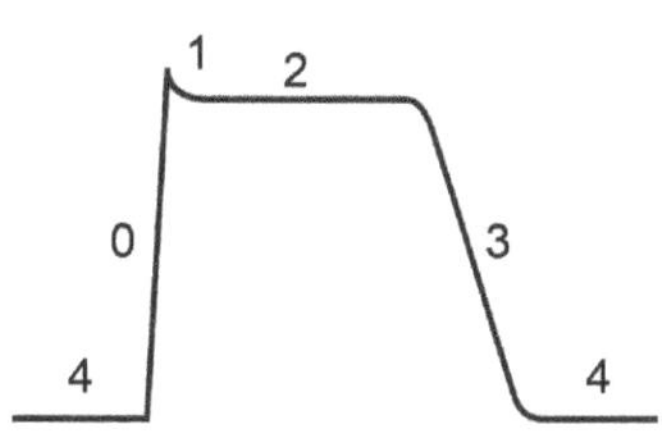

FIGURE 6.3

Phase 4: Resting membrane potential of yin and yang ion balance with Cl^- ions, a high number of K^+ ions, and a lower number of Na^+ and Ca^{2+} ions with a depolarizing voltage from −90 mV to −70 mV.
Phase 0: Rapid depolarization is caused by an increase in the electrical conduction of the fast Na^+ channels while there is a decrease in K^+ ions.
Phase 1: Inactivation of the Na^+ channels and opening of the transient outward K^+ channels.

Phase 2: The action potential is in a plateau phase with delayed repolarization. Ca^{2+} is slowly diffusing inside the cell as K^+ is quickly moving out. These L-type calcium channels open when the cell membrane depolarizes to −40 mV. They are commonly found in cardiac and vascular cells and calcium-channel blockers are usually used to block these particular channels.

Phase 3: Non-pacemaker action potential conduction quality is determined by the K^+ and Na^+ ion quantity increase as Ca^{2+} decreases.

2. Pacemaker action potentials. These potentials are the incessant and continuous movement of qi energy within the cells of the SA node, which controls the firing rates of the AV node and conduction within the ventricles. The action potential within the SA node includes phases 0, 3 and 4. These phases include changes in Ca^{2+} and K^+ funny currents, and spontaneous repolarization.

   Phase 0: The action potential is in depolarization to a voltage of −40 mV. The increase in Ca2 + ions moving through L-type calcium channels out of the cell is at a much lower rate than K^+ movement out of the cell and Na^+ ion movement into the cell.

   Phase 3: Calcium channels are deactivated, reducing the number of ions, as potassium channels open and K^+ moves inward, contributing to repolarization which ends at about − 65 mV.

   Phase 4: In the first section of this phase, the contribution of K^+ ions to depolarization is decreased. Funny currents occur due to slow inward movement of Na^+. A small increase in Ca^{2+} through T-type channels which open at low voltage of −50 mV and are not blocked by calcium channel blocker medications. The gradual increase in Ca^{2+} ions finally settles the sequence back to phase 0.

PACEMAKER ACTIONS

Baroreceptors respond to an increase or decrease in pressure. They are located in blood vessels and especially in the specialized cells of the SA node, which is the site of manifestation and transmission of heart qi. The SA node deplores at an average rate of about 110 beats per minute (bpm). Intrinsic automaticity is involved in heart rate where the ideal resting rate may be between 50 and 70 bpm, and the rate during activity can range between 80 and 200 bpm. Vagal stimulation releases acetylcholine, which slows or stops the SA qi energy firing and transmission through the conduction system. To facilitate this period, acetycholine binds to muscarinic (M2) receptors, and this decreases cAMP through inhibitory G protein. K^+ concentration is manipulated by activating a specialized potassium channel called the K_{ACh} channel, which increases K^+ ions and conduction, especially during phase 4. This is useful in understanding the pharmaceutical medications and herbal formulas used to treat arrhythmia.

The AV node is located near the septum, or the meeting point of the corners of the atrium and the ventricles. It is a collection of specialized cells which serve to delay heart qi through the conduction system to give the atria time to contract. Finally, it distributes the qi through the His and Purkinje fibers around the ventricles so that they can also contract.

BLOOD VESSELS AND BLOOD

Heart qi assists with physiologic functioning of the heart, which circulates blood throughout the body via the blood vessels. The respiratory, endocrine, hepatobiliary, gastrointestinal and renal systems use the various types of vessel as reservoirs to excrete, exchange, and deliver products such as vital nutrients, fluids, hormones, other biochemicals and their waste into the systemic circulation. These products comprise the measured composition, volume and quality of blood.

Blood is considered yin. Blood includes both yin and yang components. Yin components include water and other fluids, and yang components include oxygen, carbon dioxide, hormones, nutrients, biochemicals and microbes. Oxygen is one of the most important components within blood as it is needed by all cells and tissues throughout the body. Blood returning to the heart enters through the inferior and superior vena cavae. This deoxygenated blood also includes a variety of yin and yang components which have the capability of influencing the direct functioning of the heart. For example, chemical components from the renin–angiotensin system of the lungs and kidneys are involved in the regulation of fluid volume and blood pressure, especially within the ventricles.

After the returning systemic blood enters the right atrium, it is pumped into the right ventricle and sent through the pulmonary artery to the lungs. During gas exchange, some of the products remain in the oxygenated blood. As the blood flows through the pulmonary veins and re-enters the left heart, there is preparation for compensation and adaptation of pressure changes, stroke volume and potential cardiac output. The left ventricle will be most affected by these changes. From the left ventricle the blood then exits the endocardium through two major arteries. The cardiac vessels receive oxygenated blood to nourish the heart, and the aorta moves oxygenated blood back into the systemic circulation to repeat the holistic cycle of the heart.

The dilation and constriction of the blood vessels are also dominated by heart qi.

The vascular system delivers oxygen to all the cells of tissues and organs in the body.

In Chinese medicine, the physiologic aspects of the heart are described with the functions of governing blood and dominating blood vessels. The heart, other organs and organ systems are nourished by the yin fluids and yang products such as oxygen, which is found within the circulating blood.

The heart receives its own supply of blood from the coronary arteries. Two major coronary arteries branch off from the aorta near the point where the aorta and the left ventricle meet. These arteries and their branches supply all parts of the heart muscle with blood.

See Figures 6.4 and 6.5.

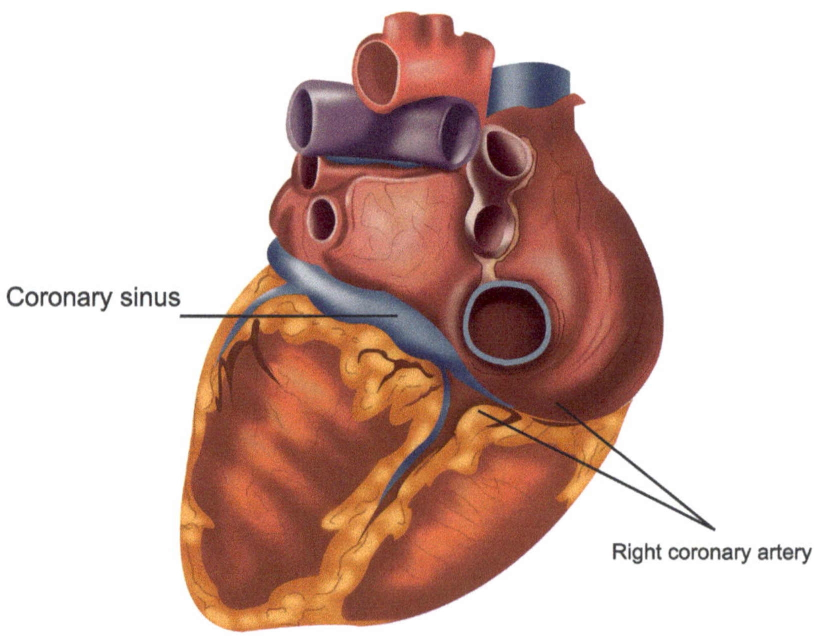

FIGURES 6.4

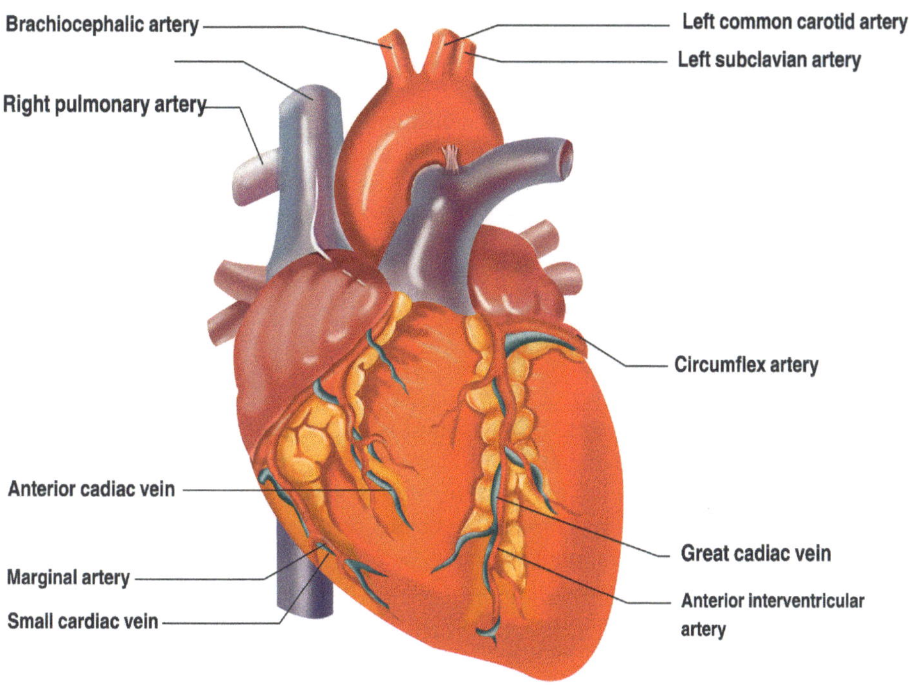

FIGURE 6.5

Right Coronary Artery (RCA)

The right coronary artery provides blood to the right atrium and ventricle, the bottom of the left ventricle, and then the septum.

Branches

Right marginal artery and posterior descending artery (PDA).

Left Coronary Arteries (LCAs)

The left coronary arteries supply blood to the left atrium, the side and the back of the left ventricle. The left anterior descending artery (LAD) supplies the front of the septum and the bottom and front of the left ventricle.

Branches

Circumflex artery and left anterior descending artery (LAD).

FACTORS IN HEART PATHOLOGY

Ischemia and Chest bi

In Chinese medicine, ischemia is considered an obstruction. Blood flow is decreased or blocked within the coronary arteries, reducing oxygen supply to the muscle tissues and causing problems with the conduction system. For example, a high homocysteine level may cause endothelial damage and inflammation of the blood vessels. Damage and inflammation lead to atherosclerosis which blocks blood and oxygen to the heart.

Inflammation

In Western medicine, acute inflammation is characterized by marked vascular changes caused by inflammatory mediators connected with the immune system. Chronic inflammation is noticed in obesity as an immune response to various microbial infections. In Chinese medicine, acute and chronic inflammatory responses are realized through the theory of pathomechanism and syndrome differentiation. Acute responses involve the vital qi being overwhelmed by any of the six pathogenic factors of wind, cold, summer heat, dampness, dryness and fire. Chronic response symptoms are explained through syndromes of dampness and phlegm, which are caused by imbalances between the liver and gall bladder, liver and spleen, and spleen and stomach.

1 Acute Inflammation

In Chinese medicine, the six pathogenic factors include six stages experienced on four levels. The stage 1 wei level includes the exterior syndromes related to cold and flu. They include heat syndromes of wind heat, damp heat, summer heat and dry heat. The stage 2 qi level is marked by the pathogen attacking and depleting the wei qi, and migrating deeper to infect the respiratory and gastrointestinal tract. They include pathogenic heat and summer heat syndromes. This stage includes the five big factors: fever, pulse, sweat, thirst and irritability. The stage 3 ying level is marked by the pathogen migrating deeper towards the heart and pericardium. Illnesses are exacerbated at night with insomnia, fever and mental disorders, and during the day with irritability, exhaustion, dry mouth and thirst. This differentiation describes pathogenic heat attack, which disturbs the balance of the heart. The stage 4 xue level is marked by the pathogen moving on to affect the blood level and quality. This also has a serious affect on the heart. The symptoms include bleeding disorders from various orifices, mania and mental disorders, blood deficiency and blood heat.

In Western medicine, acute inflammation is an immediate and temporary reaction caused by tissue injury and by infection from various pathogens. Symptoms can include:

- High fever and redness, especially at the site of the injury
- Immediate increase in heart rate, blood pressure and possible palpitations
- Appetite problems
- Sleep problems, with tiredness and lowered vitality

The immune response to acute inflammation involves leukocytes, plasma-derived mediators and complement systems.

In leukocytes, basophils respond to inflamed tissue, monocyte excess is seen in severe infection and chronic inflammation, and eosinophils and neutrophils respond to parasites and helminthes.

The plasma-derived mediators include the kinin system, the fibrinolysis system and the coagulation system. In the kinin system, bradykinin is a protein which induces the pain response, contracts smooth muscle, increases vascular permeability, causes vasodilation and increases vascular permeability. The complement system is part of the innate immune system. It involves plasmin and Factor XII. When Factor XII is activated, it initiates the kinin system, the fibrinolysis system and the coagulation system.

The complement system includes a series of complement proteins: C3, C5, C6, C7, C8 and C9. These proteins function in agglutination, and in ospinization of antigens through phagocytosis and cell lysis. Some proteins work with the coagulation system to attract leukocytes, through chemotaxis, in various parasite invasions, and in killing invading microbes through cell lysis. Among complement proteins, the cleavage of C3 produces C3a and C3b. C3a is involved with cell-derived mediators that cause vasodilation, and C3b marks bacteria for phagocytosis. The kinin system is involved in several effects of inflammation, such as vasodilation. C5 produces histamine by stimulating vasodilation and directs other complement proteins – C6, C7, C8 and C9 – to enter microbial walls and cause cell lysis. The coagulation system involves a clotting cascade. Thrombin breaks down fibrinogen and produces fibrin – which is involved in the formation of blood clots and in other inflammatory responses.

Cell-Derived Mediators

Granulocytes act as inflammatory mediators by breaking down plasma proteins. Histamine, produced in an inflammatory response, causes vasodilation and increased vascular permeability. Prostaglandins are lipids which cause vasodilation and also the fever and pain experienced during the inflammatory response.

2 Chronic Inflammation

Chronic inflammation is the long-term result of persistent infection or immune response after acute inflammation. In Western medicine, chronic inflammation, when noticed in patients with cardiovascular disease, may be associated with obesity. Especially in obesity, an elevated C-reactive protein (CRP) is possible. CRP is usually activated in chronic inflammation found anywhere in the body, and increases blood pressure and creates other risks for heart diseases. Angiopoietin (Angpt12) is another protein found in fat tissues, and high levels are associated with the development of insulin resistance, leptin resistance and chronic inflammation.

In Chinese medicine, dampness and phlegm are pathogenic factors related to hyperlipidemia and obesity. Kidney deficiency is considered the ben (root), and the bao (branch) is primarily qi stagnation and blood stasis, turbid phlegm and dampness. Other syndrome differentiations include spleen and kidney yang deficiency, liver and kidney yin deficiency.

Qi Stagnation and Blood Stasis

Symptoms: headache, neck and shoulder pain, fixed location pain, dizziness, numbness of the limbs.
Tongue: red, dark spots, veins under the tongue.
Pulse: choppy and wiry.

Turbid Phlegm

Symptoms: abdominal distention, nausea, indigestion, obesity, sensation of heaviness in the head and body, dizziness.
Tongue: thick greasy coating.
Pulse: wiry.

Spleen and Kidney Deficiency

Symptoms: coldness sensation, pale skin, edema, abdominal distention, poor appetite, dizziness, depression, low energy and poor memory.
Tongue: white and greasy coating.
Pulse: thready and deep.

Liver and Kidney Yin Deficiency

Symptoms: headache, weakness and soreness of low back and knees, dizziness, poor memory, low energy, tinnitus, insomnia, heat sensation in the palms of hands and soles of the feet.
Tongue: red with scanty coating.
Pulse: thready and rapid.

RELATIONSHIPS BETWEEN THE HEART ZANG AND OTHER ZANG AND FU ORGANS THROUGH THE VASCULAR SYSTEM

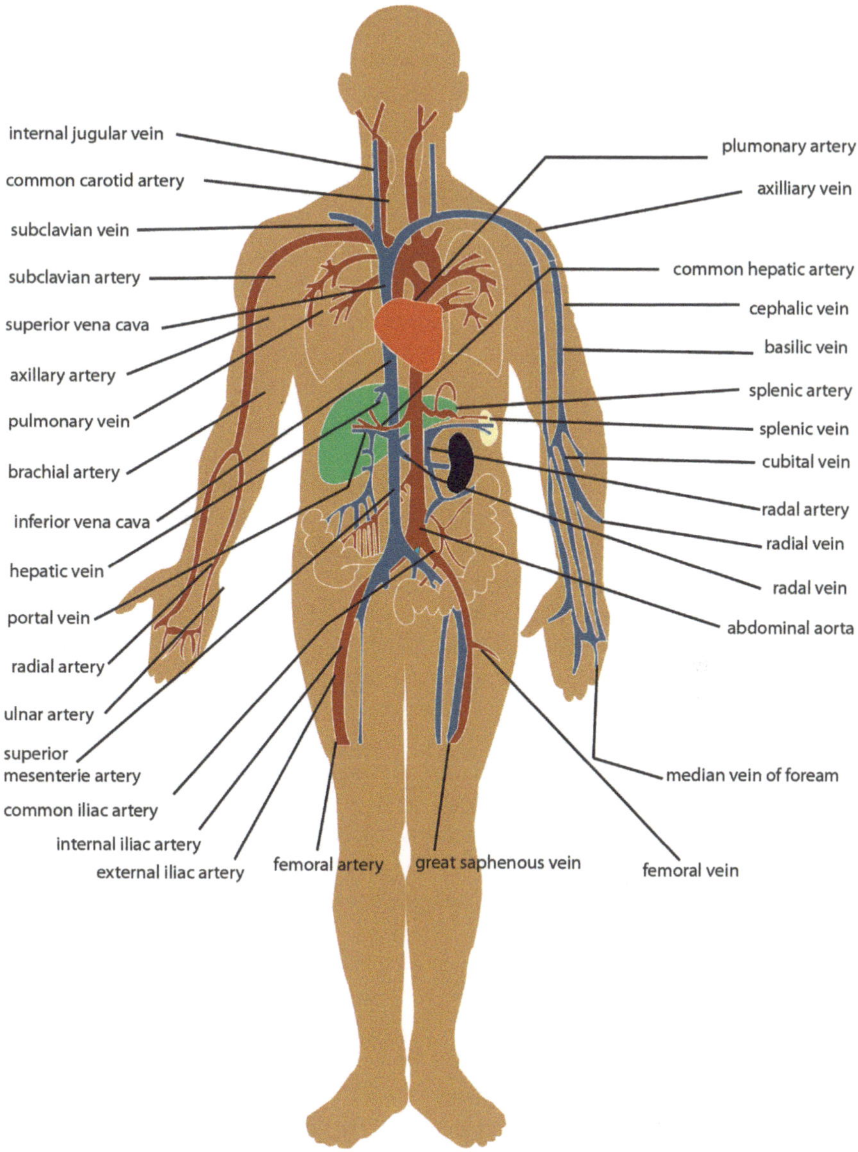

FIGURE 6.6

The Heart Zang and Small Intestine Fu

Blood flow to the intestine allows the absorption of water and nutrients from the digested food. The superior mesenteric artery is one of the three major arteries which branch from the abdominal aorta, and it connects the small intestine to the heart. Within the small intestine, branches from the artery separate into arterioles which penetrate the submucosa to supply blood and nutrients to the villi. The small intestine receives a large portion of blood from the cardiac output for the digestion of food. Fasting blood flow is around 300 ml/minute; it increases to around 900 ml/minute during meal times. Gastrointestinal bioproducts from the stomach and pancreas – such as amino acids, glucose and hormones (e.g. cholesystokinin and gastrin) – are involved in the stimulation of blood flow. In pathology, the heat from the blazing heart fire consumes blood and reduces cardiac output. The intense heat and possibly the diminishing blood flow descends and circulates through the superior mesenteric artery to bring fire to the small intestine. The signs are seen on the tongue, with sore geographic patches, and hot scanty urine which is dark in color.

The Heart Zang and Spleen Zang

The spleen is part of the functioning of the immune system. It has an abundance of monocytes for phagocytosis, as well as a blood reservoir and filter. The splenic artery is the vessel which connects the spleen to the heart. The spleen is involved in hematopoiesis, and the storing of blood. It is also involved in producing opsonins which activate the complement system – this is connected with vasodilation and phagocytosis. The splenic artery branches with the dorsal and greater pancreatic arteries, short gastric arteries, and the left gastro-omenal artery; this allows a gastrointestinal connection, with the production of blood necessary for body functioning and digestion. Stress can have an effect on the circulating blood volume. In TCM pathology, the deficiency of the spleen leads to blood deficiency; this affects the heart's ability to control a sufficient level of blood through the vessels. Weak or deficient heart qi can fail to circulate the blood – leading to blood stasis. Deficient spleen qi fails to control the blood – leading to bleeding disorders.

The Heart Zang and Lung Zang

The respiratory system is involved in two separate circulations. One is bronchial circulation and the other is pulmonary circulation. In bronchial circulation, the respiratory structures receive blood and nutrients. In pulmonary circulation, the lungs are involved in gas exchange, i.e., removing CO_2 from the blood, and replenishing O_2 in the systemic blood circulation. The lungs and heart are connected through the return of oxygen-poor systemic blood to the lungs, through the pulmonary artery, and the return of oxygen-rich blood to the heart through the pulmonary veins. All of the cardiac output (100%) moves through the right ventricle to the lungs. In TCM pathology, insufficient lung qi and pathogenic invasion lead to the consumption of blood. Weak or deficient heart qi fails to smooth the flow of blood through the pulmonary artery, which contributes to chest obstruction and breathing difficulties.

The Heart Zang and Kidney Zang

The renal system filters the blood and forms the urine. There are three major functions: (i) the cortex has an outer layer of glomeruli for filtration, (ii) the medulla includes tubules and capillaries for concentrating urine, and (iii) the hilum is concerned with the flow from the lymphatic system and blood vessels, and the exiting of urine through the ureters. The kidneys and heart are connected through the renal artery. The kidneys are involved with autoregulation of glomerular filtration through most blood pressure fluctuations, and the regulation of blood pressure. Some 20% of cardiac output influences kidney functioning, while the kidneys have a major role in vasoconstriction. Hormones from the angiotensin–renin system – as well as the aldosterone from the adrenal cortex – are involved in blood pressure rises which contribute to heat failure. In TCM pathology, if the heart fire fails to descend to warm the kidneys, and the kidney water fails to ascend to control the heart fire, then there will be disharmony between the heart and kidney. Symptoms may include insomnia, palpitations and edema. If there is a failure of the spirit to control essence, and of the essence to generate the spirit, the symptoms may include nocturnal emission, sexual problems, difficulty in concentration and thinking, and diminished intelligence.

The Heart Zang and Liver Zang

The hepatic system detoxifies blood, metabolizes glucose, and synthesizes proteins, and is a blood reservoir. The liver and heart are connected through the hepatic artery – which also receives 75% of the blood flow from the spleen, pancreas and stomach. The sinusoids within the liver are formed from the terminal vessels of the hepatic artery and hepatic portal vein, and they also function as capillaries with a very low pressure of about 5–10 mmHg. Because the central venous pressure (CVP) – which is also the right atrial pressure – shares the same low pressure, increases or changes in pressure have a direct affect upon both. Failure of the right ventricle leads to increased pressure within the sinusoids, causing hepatic and abdominal edema. In TCM pathology, the heart and liver are closely dependent upon each other with regard to the circulation of the blood, so insufficient amounts of blood, or stasis, can cause failure in physiologic functioning. If the heart blood is deficient and the liver fails to smooth the free flow of qi, then emotional and psychologic issues will appear – with symptoms such as depression, irritability and hair-trigger anger.

The Heart Zang and Brain

The brain functions as the central controller of all body functions as well as cognitive and emotional reasoning and responses. The brain is connected to the heart through four arteries: the right and left cerebral arteries, and the right and left carotid arteries, which connect to form the basilar artery. The circle of Willis regulates flow throughout the brain. It is formed by the basilar artery connection with the brainstem which connects to the carotid arteries. The brain, like the coronary circulation, requires around 20% of the oxygen circulation found in systemic blood. Moreover, the brain stem is involved in neurohumeral control. Autoregulation includes baroreception and chemoreceptors as well as autonomic and humeral control. Baroreception involves arterial blood pressure, and is regulated by baroreceptors in the carotid sinuses and the arch of the aorta. Nerve IX innervates with the carotid sinus. Chemoreceptors are cells located on the arteries and in the medulla oblongata. They regulate partial pressure of carbon dioxide (PCO_2) and affect the respiratory gas exchange of carbon dioxide and oxygen.

The autonomic nervous system includes parasympathetic control concerning the right and left vagus branch innervation, respectively, with the SA and AV nodes, other sections of the vascular system, and other organs/viscera. The sympathetic control system innervates the vessels. Humeral control is connected with the sympathetic nervous system and involves many hormones from various sources throughout the body which influence heart functioning. A circulating catecholamine, norepinephrine, couples with other adrenoreceptors to cause vasoconstriction and smooth muscle contraction. In TCM, the heart has the psychologic function of housing the inner spirit, which is associated with mental activity and the emotion of joy. Inner spirit is a product of the complex yin and yang balance of activities within the brain. Physiologic brain activity is associated more with yin because of the presence and expression of biochemicals. Cognitive functioning involves thought and emotion and is therefore immaterial and associated more with yang. Inner spirit controls heart qi through the balance of yang. In pathology, thinking is clouded and slow, the spirit lacks ground, responses to situations are irrational, and consciousness is low and dark.

THE KEY SYMPTOMS OF CARDIOVASCULAR DISEASE

Chest Pain

In Chinese medicine, chest pain is called xiong (she-yong) tong. The main concerns of xiong bi are whether the pain is associated with angina or a myocardial infarction. Angina is related to coronary artery disease, which includes poor oxygen circulation and the accumulation of plaque from blood lipids.

There are two types: stable and unstable. The symptoms include substernal pain, which radiates to the jaw, neck, both shoulders and upper back, and the left arm, and which may cause nausea.

Stable angina is chest pain which occurs in a typical and repeatable pattern during exertion. It is usually caused by walking or running up an incline, cold weather, and psycho-emotional stress. Stable angina is relieved by rest with sublingual nitroglycerine.

Unstable angina occurs with increased frequency at random times when the body is at rest.

Myocardial infarction is more intense and includes nausea and vomiting, dyspnea, sweating and palpitations. Angina, myocardial infarction and other causes in the differential diagnosis may or may not be life-threatening. However, episodes usually require immediate medical attention by a first responder and, later, in a hospital facility. Other differential diagnoses include pericarditis, esophageal spasm, gastrointestinal disorders, aortic dissection and aortic stenosis. In some cases, the causes may be due to a psycho-emotional episode or disorder.

Dyspnea

In Chinese medicine, dyspnea is called qi chuan (choo-an); it is characteristic of breathing which is uncomfortable and labored. In some cases the breathing difficulty may be related to hormonal problems, paralysis of the diaphragm, neurologic disorders, psycho-emotional disorders, and/or anemia.

There are two factors involved: cardiac and pulmonary. When the causes are related to cardiovascular diseases, the heart valve dysfunction causes regurgitation of blood and stenosis. Other concerns include congestive heart failure, arrhythmia, ischemia and cardiomyopathy. There is usually increased left atrial and ventricular pressure, where a build-up of fluids interferes with gas exchange in the lungs.

Pulmonary factors include the structures of the respiratory system as they are related to various infectious diseases, COPD, asthma, pneumothorax, pleural effusion, pulmonary embolism, etc. There are three types of cardiovascular-related dyspnea: dyspnea, paroxysmal nocturnal dyspnea and orthopnea. In dyspnea, the breathing occurs at rest. Paroxysmal nocturnal dyspnea is seen in left heart failure; the breathing is labored at night, during sleep, and it requires the patient to sit up in bed. In orthopnea the breathing is an indication of pulmonary diseases.

Palpitations

In Chinese medicine, palpitations are called xin ji (sheen-jee) and are characterized by an awareness of the rate and rhythm of heartbeats. The symptoms include skipping, fluttering, rapid tapping and pounding against the chest wall.

In addition, there may be dizziness, dyspnea, chest pain, and a light-headed sensation. Arrythmias are usually associated with palpitations with specific patterns:

- Abrupt onset: supraventricular tachycardia and ventricular tachycardia
- Abrupt onset with syncope: ventricular tachycardia
- Missed beats: premature ventricular contractions
- Slow-paced rhythm: sinus bradycardia and heart block
- Rapid and regularly paced rhythm: supraventricular tachycardia and ventricular tachycardia
- Rapid and regular with palpitations and neck pulse sensation: atrial ventricular node re-entrant tachycardia
- Rapid and irregularly paced rhythm: atrial fibrillation

PULSE QUALITIES

Strong Pulse

Chinese medicine: interior heat, toxic heat, yin deficiency, yang excess, fire, deficient qi.
Western medicine: fever, pain, inflammation, infection, hypertension, arrhythmia, sick sinus syndrome, atrial flutter, tachycardia, anemia, dehydration, ventricular fibrillation, viral myocarditis.

Wiry pulse
Smooth pulse

Chinese medicine: dampness, phlegm obstruction, yang rising, qi stagnation, deficient spleen and kidney yang.
Western medicine: hyperproteinemia, pain, psychosis, pregnancy, cerebral thrombosis.

Irregular pulse
Pulsus bigeminus
Pulsus alterans
Pulsus bisferiens

Chinese medicine: qi deficiency, yin deficiency, blood stasis.
Western medicine: premature ventricular contractions, infection, myocardial infarction, cardiomyopathy, aortic stenosis, angina pectoris, hypertension.

Choppy pulse

Chinese medicine: blood deficiency.
Western medicine: anemia.

Taut pulse

Chinese medicine: maladjustment of chong and ren meridians.
Western medicine: hypertension.

Thready pulse

Chinese medicine: maladjustment of chong and ren meridians, kidney deficiency.
Western medicine: rheumatic heart fever.

Weak/slow pulse

Chinese medicine: fluid retention, yang deficiency, cold accumulation, blood stasis, heart and kidney yang deficiency, qi deficiency.
Western medicine: sick sinus syndrome, bradycardia, syncope, thromboangiitis obliterans, viral myocarditis, cerebral thrombosis, Takayasu arteritis.

Absent pulse
Pulsus paradoxus

Chinese medicine: blood stasis, damp accumulation, separation of yin and yang
Western medicine: cardiac or pulmonary effusion, COPD, pneumothorax, pericarditis, right ventricular obstruction, myocardial infarction, shock, vascular obstruction, death.

Scattered pulse
Intermittent pulse
Absent pulse
Slow pulse

EVALUATING CARDIOVASCULAR DISEASE

The clinical context for cardiovascular diseases in the integrative Chinese medicine setting is important. In Western medicine, the differential diagnosis points towards the causes of certain symptoms and the extent to which those symptoms are related to cardiovascular decline. In Chinese medicine, the main idea is about the magnitude of yin and yang imbalance towards or during cardiovascular decline. The concern is for prevention or management of heart failure, valvular and coronary artery diseases.

While taking a health history, the health practitioner should look for clues linking symptoms to genetic, dietary and lifestyle causes. For example, vascular problems can be noticed in claudication, edema and skin discolorations. Claudication can be seen while the patient travels to the exam room in a clinic and while they mount the exam table. Breathing difficulties can accompany fatigue during ordinary movement.

Listening to the patient's narrative should give insight into their thoughts, attitudes, fears, prejudices and preferences.

Physical examination should include mainly a Western medicine examination of the precordium, pulses, auscultation, imaging and laboratory studies. Tongue and pulse diagnosis in Chinese medicine is used mainly to give indications for physiologic reactions to treatment.

SUMMARY IN THE CONTEXT OF CHINESE MEDICINE AND WESTERN MEDICINE SYSTEMS IN PERSONALIZED MEDICINE

The heart is considered to be the monarch of all zang and fu organs in the Chinese medicine system. Pathologic factors between heart and zang/fu organs include the direct relationship through the connecting arteries. Organs are affected according to various pressures within the heart, sufficiency of blood and oxygen. Among several, the key symptoms of cardiovascular disease include chest pain, dyspnea and palpitations. These symptoms determine the indication and contraindications involved while considering integrating Chinese medicine and Western medicine.

Part of the initiation or exacerbation of cardiovascular diseases includes ischemia and inflammation. In Western medicine, acute inflammation is characterized by marked vascular changes caused by inflammatory mediators connected with the immune system. Chronic inflammation is noticed in obesity as an immune response to various microbial infections. In Chinese medicine, it is caused by the six pathogenic factors and syndromes of dampness and phlegm, which include vascular changes caused by inflammatory mediators with the immune system. Ischemia is called xiong bi (shee-ong bee) or chest bi in Chinese medicine. Obstructed blood flow in the coronary arteries reduces blood supply and also causes conduction problems.

For providing personalized medicine, it is important for health practitioners to establish theories, sciences, examination methods, diagnoses and treatment methods which are common and distinct. The target goal in treatment is for the patient to live a life free of need for medications to treat the body based on cardiovascular disease symptoms. Generally, the goal is to make the use of Chinese medicine a part of the prevention strategy towards the target. Further extremes in health, which necessitate temporary pharmaceutical intervention, require discernment over the appropriateness of integrating Chinese herbal medicines with pharmaceutical drugs. When appropriate, Chinese medicine can act as a protector of organ and immune systems during treatment, for weaning off pharmaceutical drugs at the end of a course of treatment, and to stabilize a planned prevention program.

FURTHER READING

[1] Kefu C, Qingrong Z. Fundamental theory of traditional Chinese medicine. 2nd ed. Beijing, China: People's Medical Publishing House; 1998.

[2] Bonow RO, Mann DL, Zipes DP, Libby P. Braunwald's heart disease: a textbook of cardiovascular medicine. 9th ed. Philadelphia, PA: Saunders Elsevier; 2011.

[3] Noble A, Johnson R, Thomas A, Bass P. The cardiovascular system: basic science and clinical conditions. Singapore: Elsevier Singapore; 2010 Peking University Medical Press

[4] Lu A, Jiang M, Zhang C, Chan K. An integrative approach of linking traditional Chinese medicine pattern classification and biomedicine diagnosis. J Ethnopharmacol 2012;141(2): 549–56.

[5] Ma T, Tan C, Zhang H, et al. Bridging the gap between traditional Chinese medicine and systems biology: the connection of Cold Syndrome and NEI network. Mol BioSystems 2010;6(4):613–9.
[6] Lu OP, Chen KJ. Integrative medicine in clinical practice: from pattern differentiation in traditional Chinese medicine to disease treatment. Chin J Integr Med 2009;15(2):152.
[7] Lu P, Chen KJ. Chinese medicine pattern diagnosis could lead to innovation in medical sciences. Chin J Integr Med 2011;17:811–7.
[8] Paul UU. Huang Di Nei Jing Su Wen: nature, knowledge, imagery in an Ancient Chinese medical text. Los Angeles, Calif, USA: University of California Press; 2003.
[9] Veith I. Huang Ti Nei Ching Su Wen. The Yellow Emperor's classic of internal medicine. Baltimore, Maryland, USA: Williams & Wilkins; 1949.
[10] Harmsworth K, Lewith GT. Attitudes to traditional Chinese medicine amongst Western trained doctors in the People's Republic of China. Soc Sci Med 2001;52:149–53.
[11] Wang S. On the modalities of modern traditional Chinese medicine education. J Shanxi Coll Tradit Chin Med 2012;3:69–71.
[12] Fushimi R, Tachi J, Amino N, Miyai K. Chinese medicine interfering with digoxin immunoassays. Lancet 1989;333:339.
[13] Wang J. Evidence-based medicine in China. Lancet 2010;375:532–3.
[14] Zhang L, Yan J, Liu X, et al. Pharmacovigilance practice and risk control of Traditional Chinese Medicine drugs in China: current status and future perspective. J Ethnopharmacol 2012;140:519–25.
[15] Stone R. Biochemistry. Lifting the veil on traditional Chinese medicine. Science 2008;319:709–10.
[16] van der Greef J, Hankemeier T, McBurney RN. Metabolomics-based systems biology and personalized medicine: moving towards n = 1 clinical trials? Pharmacogenomics 2006; 7:1087–94.
[17] Uzuner H, Bauer R, Fan TP, et al. Traditional Chinese medicine research in the post-genomic era: good practice, priorities, challenges and opportunities. J Ethnopharmacol 2012;140:458–68.
[18] Xu Q, et al. The quest for modernisation of traditional Chinese medicine. BMC Complement Altern Med 2013;13:132.
[19] Jiang M, Zhang C, Zheng G, et al. Traditional Chinese medicine Zheng in the era of evidence-based medicine: a literature analysis. Evid Based Complement Alternat Med 2012;409568.
[20] Maximilian Buja L, Entman ML. Modes of myocardial cell injury and cell death in ischemic heart disease. Am Heart Assoc 1998;10.1161/01.CIR.98.14.1355.
[21] Amri H, Haramati A. Using basic science to develop an innovative program in complementary and alternative medicine. J Int Assoc Med Sci Educ 2010;20(2):48–55.

Chapter 7

Personalizing Medicine

Chapter Objectives

After studying this chapter you should be able to:

1. Explain the challenges of integrating aspects of personalized medicine for cardiovascular care
2. Explain the benefits of integrating aspects of personalized medicine for cardiovascular care
3. Describe what information genome-wide association studies (GWAS) provide for cardiovascular predisposition, diagnosing and treatment
4. Discuss how constitutional theory helps in forming personalized prevention and treatment strategies
5. Summarize the context of the patient role in personalizing medicine

OVERVIEW

This chapter outlines the main aspects of personalizing integrative Chinese medicine and Western medicine. In Chinese medicine, constitutional theory is used for profiling patients, diagnosing and treating. In Western medicine, genome-wide association studies (GWAS) are used to associate genes with certain congenital factors that predispose towards certain cardiovascular diseases, diagnosis and management using pharmaceutical drugs.

CONSTITUTIONAL THEORY

(a) General and modifiable treatment suggestions according to the individual patient:
- Male
- Female
- Infant and childhood constitution
- Teenage and young adult
- Middle age
- Geriatric

Integrative Cardiovascular Chinese Medicine. DOI: http://dx.doi.org/10.1016/B978-0-12-420014-2.00005-3

(b) General and modifiable treatment suggestions according to season:
- Spring to summer
- Summer heat
- Fall (autumn)
- Winter

(c) General and modifiable cardiovascular disease prevention suggestions:
- Regulate qi and blood
- Eliminate excess patterns in the spring
- Fall (autumn) balancing
- Winter stabilizing
- Spring balancing
- Re-supply any deficiencies after recovery from serious illness
- Purge and dispel exogenous pathogens
- Invigorate the viscera

(d) General and modifiable cardiovascular disease treatment suggestions based on routine root treatment.

(e) General and modifiable cardiovascular disease treatment suggestions based on emergency branch treatment.

DIAGNOSIS AND TREATMENT ACCORDING TO GENOME

1. Genetics Associated with Cardiovascular Disease:

- 4q25
- miR-376c
- *MYH9* gene
- 5-gene profile
- Familial hypercholesterolemia (FH)
- Sodium channel *SGN5A* gene, *KCN* (potassium channel) gene
- HLA-DR7 (human leukocyte antigen, HLA)
- HLA-DQA1
- CFHR1 and CFH3
- *MS4A1* gene
- Interleukins IL-6, IL-12B, IL-23
- Interleukins IL-6, IL-12B
- HLA-B and HLA-D
- *FCGR2A*/*FCGR3A* and *PSMG1*

2. Clinical Uses for Genetics in Cardiovascular Diagnostic Testing:

- *HCN4* gene
- *DKN2A* and *DKN2B*
- *MTAP*
- Factor V and prothrombin genes

3. Clinical Uses for Pharmacogenomics in Selection of Cardiovascular Medications:

- Cytochrome P450
- Genotype (*CYP2C9*)
- *CYP4F2*
- Vitamin K epoxide reductase complex genotype 1 (*VKORC1*)
- *R506Q* and *20210G*
- Metabolizing liver enzyme: *CYP2C19*
- Paraoxonase I (*PONI*)
- *P2Y12*
- Paraoxonase I (*PONI*) genotype
- *SLCO 1B1*
- *KIF6* gene
- Bradykinin type I (BKI) receptor haplotype
- Angiotensin II (AT-II) type I receptor haplotype
- Apolipoprotein A5 (ApoA5) genotype
- Niemann–Pick CI-like I (*NPCILI*) haplotype

Key Points

- Realizing the variations among individuals based on specific genes may help to determine the susceptibility to diseases. Also, it can guide prevention practices.
- Although realizing gene variations among individuals may help to determine disease susceptibility to diseases, though it may not always determine the responses to treatment or successes.
- Incorporating genomic studies and Western medicine methods into Chinese medicine for clinical use rests mainly on the discretion of the individual health practitioner.
- Genomic information can be used in the clinic for cardiovascular disease identification, diagnostic testing and pharmacogenomic consulting.
- GWAS have provided important information which has helped medical practitioners to recognize disease risk factors and possible drug metabolism problems, and has aided the development of certain drugs which target the individual needs of patients.
- Constitutional treatment according to the individual patient should be modified accordingly.
- General and modifiable treatment according to season is strongly suggested.
- General and modifiable cardiovascular disease prevention suggestions include regulating qi and blood, eliminating excess patterns in the spring, balancing in the fall (autumn) to prepare for winter, stabilizing in the winter, re-balancing in the spring to prepare for the fall (autumn), supplying deficiencies after illness, purging and dispelling pathogens to prevent

cardiovascular and blood complications, invigorating the viscera and harmonizing the relationship between them.

- The root is considered the primary cause of disease.
- The branch is considered any temporary complication or secondary and indirect condition.

INTRODUCTION

In both Chinese medicine and Western medicine, it is common and regular practice to personalize preventive and therapeutic methods to treat the person as well as to address the disease. About a thousand years ago, Hippocrates stated that "It's far more important to know what person the disease has than what disease the person has". This statement is important to keep in mind for maintaining a clinical perspective and being of service to patients. Personalized medicine has expanded to include more tools, such as genetic and other biomarkers. These tools are becoming quite important in Western medicine practice, where bringing more humanity into the examination room helps to put the patient's identity more into focus. It is worth taking a look at how Chinese medicine views the person first and determines the treatment principle and treatment on a regular basis.

Sir William Osler once noted that "If it were not for the great variability among individuals, medicine might well have been a science and not an art". In China, cardiovascular departments of all Chinese medicine hospitals include the use of allopathic procedures, pharmaceutical drugs singly or in combination with herbs and other substances in tailored formulas. Patients can visit the outpatient clinic and also be admitted as an inpatient, and receive intravenous herbs and pharmaceutical drugs.

PERSONALIZING INTEGRATIVE CHINESE AND WESTERN MEDICINE

Personalized medicine using genome studies is already being practiced in cardiology, with continued modifications which help with future standardizations. For now, research on integrative Chinese medicine with Western medicine on cardiovascular patient samples with a variety of genetic variants remains low to non-existent. Incorporating genomic studies and Western medicine methods into Chinese medicine for clinical use rests mostly on the discretion of the individual health practitioner. There are three clinical issues to overcome:

- Practitioners must be willing to move into the era of genomics and use risk markers or imaging in Chinese medicine, if it will benefit the patient beyond tongue and pulse diagnosing. How genomics benefits the patient

in Chinese medicine may depend on whether the practitioner has become involved at prevention, recovery/remission or end stage.
- Practitioners must be willing to incorporate biomarker information of an individual into a specific subgroup which can enhance the selection of herbal components for a decoction.
- Current research on Chinese medicine cardiology is still limited to China. It includes mostly non-standardized data collection procedures on certain herbs and modifiable decoctions. In addition, studies using female subjects and those of other races and ethnicities remain low to non-existent.

In Western countries where Chinese medicine is establishing a niche, some practitioners realize that integrating systems poses both challenges and benefits.

Challenges

Realizing the variations among individuals based on specific genes may help to determine the susceptibility to diseases. It may not always help determine the responses to treatment or successes.

- Limitations in the ability of the health practitioner, approaches in medicine and health care, and choices of therapy.
- The genetic context in the natural evolution of multifactorial cardiovascular diseases is not completely understood and, consequently, clinical applicability is still limited.
- Many traditional Chinese medicine purists resist specialization in a particular subject area of Chinese medicine and the use of Western medicine devices and drugs.
- Concern at more serious or late stages of cardiovascular disease, which influences clinical decisions and may reduce personalized care. Adverse patient lifestyle choices, psycho-emotional state, environmental exposure and personal decisions which influence health, disease and recovery.
- It is difficult to determine whether genetic testing will improve or prevent clinical events or improve medical outcomes.
- Standardization through laboratory variables can be difficult to put into clinical practice owing to observer bias.
- Risk factors and diagnostic testing in European patients may be inappropriate for other races (African, Asian, etc. descent).

Benefits

Realizing the variations among individuals based on specific genes may help to determine the susceptibility to diseases. Also, it can guide prevention practices. For example, practitioners who use Chinese medicine treatment principles and methods can begin to view herbs and substances in formulas by their actions at the genetic level rather than their individual properties alone.

- Personalized medicine aims to address the individual patient.
- Personalized medicine may provide a better strategy for efficient treatment method, safety and duration.
- Sex, race and ethnic determinants may or may not be needed to determine treatment course.
- Personalized medicine may help reduce the costs of care.
- Genome-wide association studies (GWAS) can help subgroups of patients which are not often represented in published medical studies.
- Aspects of Chinese medicine such as in constitutional theory, can be adapted to provide a way to connect with the patient and assist with diagnostic and treatment strategies and lifestyle modifications.

WESTERN MEDICINE DIAGNOSIS AND TREATMENT SUGGESTIONS ACCORDING TO GENOME

GWAS have provided important information which has helped medical practitioners to recognize disease risk factors and possible drug metabolism problems, and has aided the development of certain drugs which target the individual needs of patients.

1. Clinical Uses for Genetics According to Cardiovascular Disease

(a) Arrhythmia:
- 4q25
- miR-376c

The 4q25 gene is associated with the development of atrial fibrillation. miR-376c is considered a genetic variant in black Americans which prevents the accumulation of phosphatidylcholine transfer protein (PC-TP), which is involved in thrombus formation, leading to arrhythmia.

(b) Nephritic syndrome:
- *MYH9* gene

This gene is associated with critical elevations in blood pressure leading to kidney failure.

(c) Long QT syndrome (LQTS):
- Familion® 5-gene profile

This gene profile helps to identify the congenital origins, for early detection and also for prevention.

(d) Hyperlipoproteinemia:
- Familial hypercholesterolemia (FH)

These genetic determinants include a predisposition for accumulation of triglycerides, LDL and cholesterol at specific sites in the body.

(e) Brugada syndrome, Romano–Ward syndrome:
- Sodium channel *SGN5A* gene, *KCN* (potassium channel) gene

These genes are associated with defects in the sodium and potassium channels, which can lead to several types of cardiac rhythm disorders.

(f) Autoimmune diseases:
- HLA-DR7 (human leukocyte antigen HLA)
- HLA-DQA1
- CFHR1 and CFH3
- *MS4A1* gene
- Interleukins IL-6 , IL-12B, IL-23

These protein antigens are often responsible for increasing the possibility of autoimmune reactions in some patients.

(g) Inflammation:
- Interleukins IL-6, IL-12B
- HLA-B and HLA-D
- FCGR2A/FCGR3A, and PSMG1

When activated, these contribute to reduced myocardial contractility and damage to blood vessel endothelium.

2. Clinical Uses for Genetics in Diagnostic Testing

(a) *HCN4* gene. Mutations in this gene may be involved in the abrupt switch between sinus bradycardia and tachycardia in sick sinus syndrome.
(b) *DKN2A* and *DKN2B*. These genes are involved in regulating cellular aging and lysis, and the formation of plaque within the arteries.
(c) *MTAP*. The product of this gene processes cellular wastes into methionine, which metabolizes excess homocysteine. Mutations are involved in failures in these processes.
(d) Factor V and prothrombin genes. These genes are responsible for thrombus formation in young patients.

3. Clinical Uses for Drugs According to Pharmacogenomics

Genetic variants can affect drug pharmacokinetics (absorption, distribution, metabolism, and excretion) and/or the drug pharmacodynamics (interaction and its mechanism).

(a) Warfarin:
- Cytochrome P450
- Genotype (*CYP2C9*)

- *CYP4F2*
- Vitamin K epoxide reductase complex genotype 1 (*VKORC1*)
- *R506Q* and *20210 G*

These genes can affect bleeding, hyper-coagulation and metabolism within the liver. In addition, vitamin K levels affect drug responses.

(b) Clopidogrel:

- Cytochrome P450
- Metabolizing liver enzyme: *CYP2C19*
- Paraoxonase I (*PONI*)

These genes and enzymes are involved in thrombosis formation associated with percutaneous coronary intervention, stent placement, and metabolism of this drug for clinical benefit.

(c) Aspirin:

- P2Y12
- Para-oxonase I (PONI) genotype

Both can be used for detection of platelet reactivity in relationship to thrombotic and bleeding events.

(d) Statins:

- *SLCO 1B1*
- *KIF6* gene

These genes are detected to determine the use and dosage of this drug for hypercholesterolemia.

(e) ACE inhibitors:

- Bradykinin type I (BKI) receptor haplotype
- Angiotensin II (AT-II) type I receptor haplotype

These haplotypes are detected to determine the clinical benefit of this drug.

(f) Fenofibrate:

- Apolipoprotein A5 (*ApoA5*) genotype

This genotype is detected to determine the clinical benefit of this drug.

(g) Ezetimibe:

- Niemann–Pick CI Like I (*NPCILI*) haplotype

This haplotype is detected to determine the clinical benefit of this drug.

CHINESE MEDICINE DIAGNOSING AND TREATMENT USING CONSTITUTIONAL THEORY

(a) General and modifiable treatment suggestions according to the individual patient:
- Male: regulate qi, nourish blood, essence and yin
- Female: tonify qi and yang, nourish blood
- Infant and child constitution: reinforce the exterior, nourish essence, regulate qi and blood
- Teenage and young adult: regulate essence, nourish blood, balance yin and yang
- Middle age: replenish and regulate essence, tonify qi, nourish blood, balance yin and yang
- Geriatric: reinforce the exterior, nourish essence, regulate qi and blood, balance yin and yang

(b) General and modifiable treatment suggestions according to season:
- Spring to summer: the temperature changes from cool to warm, yang rises, balance yin and yang and exercise caution giving acrid herbs to avoid injury and dispersal of yin
- Summer heat: the temperature changes from warm to hot, yang is dominant, dampness spreads, exercise caution giving yin-nourishing herbs
- Fall (autumn): the temperature changes from hot to warm to cool, yang wanes, dryness affects the dampness, include acrid herbs and cool herbs to nourish yin and moisten dryness
- Winter: the temperature changes from cool to cold, yin rises, balance yin and yang and exercise caution on cold herbs to avoid injury and collapse of yang

(c) General and modifiable cardiovascular disease prevention suggestions:
- Regulate qi and blood: this disharmony is the basis of yin and yang imbalance. Qi produces, regulates and circulates blood, and blood is the mother of qi
- Eliminate excess patterns in the spring: balance yin and yang by warming the cold and cooling the hot, reinforcing the exterior for the autumn
- Fall (autumn) balancing: stabilizing yin and tonifying kidney yang, balance heart, kidney and liver for winter
- Winter stabilizing: tonify kidney yang, tonify heart qi, subdue liver yang
- Spring balancing: stabilize
- Supplying the deficiencies after illness: nourishing yin and stabilizing yang, stabilizing yin and tonifying yang, raising both yin and yang by nourishing, tonifying and stabilizing

- Purge and dispel pathogens: relieve the exterior, ascend the clear, descend the turbid, cool the hot, warm the cold, purge heat and fire toxins, tonify qi during exterior conditions
- Invigorating the viscera: balance the harmony between the organs by invigorating the deficient organs and subduing and purging the excess in other organs

(d) General and modifiable cardiovascular disease treatment suggestions based on routine root treatment:
- The root may be considered the primary cause of disease
- The root of disease symptoms includes two parts: the immediate and temporary symptoms after affection, and the underlying essence of the disease
- The essence of the root is usually based on visceral disharmonies, yin, yang, qi, blood, fluid, phlegm, toxin, cold or heat stagnation, deficiency or excess
- Treatment is not often palliative
- Treatment is based on acute sections of the disease first and chronic later
- Seasonal review and assessment is necessary to prepare for critical summer and winter months.
- Treatment may primarily involve regulation and prevention methods to protect, nourish and smooth the flow of qi and blood
- The basic purpose is to balance yin and yang

(e) General and modifiable cardiovascular disease treatment suggestions based on emergency branch treatment:
- The branch is considered any temporary complication or secondary and indirect condition
- Branch treatment is an emergency situation which may complicate important root treatment
- Branch treatment includes the critical acute period which must be cleared to avoid life-threatening consequences with the root (ex: exogenous pathogenic invasion which may redirect the entire treatment course during antibiotic therapy)
- Monitor the root while clearing branch complications as quickly as possible. Sometimes the root treatment must continue during branch treatment.

SUMMARY IN THE CONTEXT FOR PERSONALIZING MEDICINE

Health practitioners who endeavor to integrate Chinese and Western medicine for cardiovascular care may wish to consider incorporating genomic studies in cardiology as an additional tool in modern Chinese medicine; this provides a novel method for prevention and treatment strategies and also for further research on clinical methods, individual herbs and formulas.

FURTHER READING

[1] Aspinall MG, Hamermesh RG. Realizing the promise of personalized medicine. Harv Bus Rev 2007;85: 108–117, 165.

[2] Ginsburg GS, Willard HF. Genomic and personalized medicine: foundations and applications. Transl Res 2009;154:277–87.

[3] O'Donnell CJ, Nabel EG. Cardiovascular genomics, personalized medicine, and the National Heart, Lung, and Blood Institute: part I: the beginning of an era. Circ Cardiovasc Genet 2008;1:51–7.

[4] Vasan RS, et al. Genetic variants associated with cardiac structure and function: a meta-analysis and replication of genome-wide association data. JAMA 2012;302:168–78.

[5] Shimizu W. Clinical impact of genetic studies in lethal inherited cardiac arrhythmias. Circ J 2008;72:1926–36.

[6] Aithal GP, Day CP, Kesteven PJ, Daly AK. Association of polymorphisms in the cytochrome P450 CYP2C9 with warfarin dose requirement and risk of bleeding complications. Lancet 1999;353:717–9.

[7] Muin J, Khoury MD, et al. A population perspective on how personalized medicine can improve health. Am J Prev Med 2012;42(6):639–45.

[8] Villard E, et al. A genome-wide association study identifies two loci associated with heart failure due to dilated cardiomyopathy. Eur Heart J 2011;32:1065–76.

[9] Thanassoulis G, Vasan RS. Genetic cardiovascular risk prediction: will we get there? Circulation 2010;122:2323–34.

[10] Cooper GM, et al. A genome-wide scan for common genetic variants with a large influence on warfarin maintenance dose. Blood 2008;112:1022–7.

[11] Takeuchi F, et al. A genome-wide association study confirms *VKORC1, CYP2C9*, and *CYP4F2* as principal genetic determinants of warfarin dose. PLoS Genet 2009;5: e1000433.

[12] Evans J, Khoury MJ. Evidence based medicine meets genomic medicine. Genet Med 2007;9:799–800.

[13] Link E, et al. SLCO1B1 variants and statin-induced myopathy – a genomewide study. N Engl J Med 2008;359:789–99.

[14] Exner DV, Dries DL, Domanski MJ, Cohn JN. Lesser response to angiotensin-converting enzyme inhibitor therapy in black as compared with white patients with left ventricular dysfunction. N Engl J Med 2001;344:1351–7.

[15] Shekelle PG, Rich MW, Morton SC, et al. Efficacy of angiotensin-converting enzyme inhibitors and beta-blockers in the management of left ventricular systolic dysfunction according to race, gender, and diabetic status: a meta-analysis of major clinical trials. J Am Coll Cardiol 2003;41:1529–38.

[16] Lunshof JE, Pirmohamed M, Gurwitz D. Personalized medicine: decades away? Pharmacogenomics 2006;7:237–41.

[17] van Werkum JW, et al. A head-to-head comparison between the Verify Now P2Y12 assay and light transmittance aggregometry for monitoring the individual platelet response to clopidogrel in patients undergoing elective percutaneous coronary intervention. J Thromb Haemost 2006;4(11):2516–8.

[18] Domchek SM, et al. Association of risk-reducing surgery in *BRCA1* or *BRCA2* mutation carriers with cancer risk and mortality. JAMA 2010;304:967–75.

[19] Lee M-S, Flammer AJ, Lerman LO, Lerman A. Personalized medicine in cardiovascular diseases. Korean Circ J 2012;42:583–91.

Chapter 8

Health Communications in Cardiovascular Chinese Medicine

Chapter Objectives

After studying this chapter you should be able to:

1. Discuss the risk factors for cardiovascular diseases
2. Describe the scope of care in cardiology departments of Chinese medicine hospitals in China
3. Discuss theories and models in health literacy
4. Describe the educational level and practice scope of licensed Oriental medicine practitioners across the United States
5. Summarize the context of health communications in personalizing medicine

OVERVIEW

As communities become more aware of, and comfortable with, Chinese medicine as a regular health option, health practitioners must meet certain basic public expectations which are part of health communications. These include health literacy, improving patient outcomes and becoming more of an integrative medicine practice.

In health literacy, practitioners should be prepared to use persuasive messages and factual information about how Chinese medicine is available to assist patients. This includes, but is not limited to, practitioner level of education and qualifications, continuing education requirements, clinical protocols, indications and contraindications of acupuncture treatments, formulas and herbal ingredients and their interactions with Western medicine as necessary to protect the public.

For improving patient outcomes, practitioners must be prepared to integrate Chinese medicine terminology with medical terminology, so that the medical and public communities can understand the connection which may increase

Integrative Cardiovascular Chinese Medicine. DOI: http://dx.doi.org/10.1016/B978-0-12-420014-2.00034-X

acceptance. Practitioners should conduct research and develop small clinic studies to become more aware of how cardiovascular disease affects people in different racial, cultural and genetic groups in their caseload. This information is useful to make practitioners better health care providers. Practitioners should develop an understanding of public health theories, and models and skills in public health surveillance methods, including incidence, prevalence, and risk.

To create more of an integrative medicine practice, health practitioners must prepare to expand their breadth of knowledge from Chinese medicine to other disciplines which also serve the public, including Western medicine. Also, they must correspond or team up with Western medicine practitioners to assist those patients with acute and chronic disorders who must receive the best available care.

Key Points

- Heart disease is now a leading cause of death internationally.
- Currently, the incidence of cardiovascular disease is equal in males and females worldwide.
- In China, integrative Chinese medicine and Western medicine basic theories and sciences, diagnostic methods and treatment methods are regular protocol in the cardiology department.
- Health literacy programs can encourage more public awareness and acceptance of Chinese medicine in cardiology.
- Health communications include public health theories and models.
- Support theories were developed to improve practitioner bedside manner and to assist in better patient outcomes.
- Practitioners of Chinese medicine and others must be prepared to shape public attitudes and behaviors regarding the provision of cardiovascular care.
- Practitioners of Chinese medicine study more about incidence and prevalence, causes and risk factors, and various health care methods available to help patients.

INTRODUCTION

Governments around the world have health agencies such as the Centers for Disease Control, and National Institutes of Health. Their mission is to provide public health and surveillance programs. Cardiovascular disease is treated using Western medicine methods of physical examination, diagnosis and treatment. It is studied at institutions which view only Western medicine methods and where practitioners study and practice only in Western hospitals, clinics and treatment facilities. Heart disease is a major chronic health condition in the United States. It is the leading cause of death among all racial and ethnic groups. Males and females display different symptoms of heart disease; however, there are common risk factors which increase the likelihood of developing symptoms.

Common risk factors:

- Uncontrolled hypertension
- Smoking habit
- Uncontrolled dangerous cholesterol levels

Heart disease is described in populations by its prevalence and incidence. Prevalence means how widely the disease occurs in a population. Incidence means the frequency and magnitude in which disease occurs. Between 2000 and 2010 the prevalence of heart disease changed in the United States. Before that time, more than 80% of adults between the ages of 40 and 60 had experienced one of the associated risk factors. Currently, the prevalence has decreased to around 50% for both males and females. Decreases in males have occurred more prevalently in the age groups 40–59 and 60+. For females, the decrease was found only among the 60+ age group. The reason for the disparities in females may be that heart disease is still often misdiagnosed in women. According to health agencies in 2007, 38% of adults in the United States used complementary and alternative medicine for any acute or chronic conditions. Perhaps within this figure is another reason for the decline in prevalence of cardiovascular disease.

CHINESE MEDICINE AND THE TREATMENT OF CARDIOVASCULAR DISEASE

Chinese medicine is a system of medicine similar to, and different from Western medicine. The similarities are that there are methods of studying basic medical sciences and theories, algorithms for diagnosing and guidelines for prevention, diagnostic criteria, treatment principles, treatments methods, modalities, medicines, prognosis predictions, and research. The differences include terminology, the perspective of viewing disease manifestation and state, examination and diagnostic methods, medicines and procedures in various treatments, and the prediction of outcomes.

In China, Chinese medical universities teach students basic medical theories and sciences of both Chinese medicine and Western medicine. Chinese medicine hospitals are nearly as advanced in protocols and medical procedures as Western hospitals. Cardiovascular departments in Chinese hospitals include Western medicine examination procedures, diagnostic testing and treatments. It is common to find interventional departments for cardiovascular treatment in Chinese medicine and Western hospitals, but in Chinese medicine hospitals the scope is often limited to interventional diagnosing and/or treatments.

Currently, Chinese medicine can be used in cardiovascular diseases such as coronary artery disease hypertension, heart failure, hyperlipoproteinemia, sinus rhythm disorders such as arrhythmia and sick sinus syndrome, infectious diseases which affect sections of the heart (such as viral myocarditis), rheumatic fever and endocarditis, vascular diseases (such as thromboangiitis obliterans and Takayasu arteritis), as well as certain cardiopulmonary disorders. There are currently over 200 medicinal formulas in oral, injectable and intravenous forms, utilizing over 150 herbs. Patients can receive care either as an outpatient in the clinic or as an inpatient.

HEALTH LITERACY

Much of the work on health literacy has focused on problems and solutions for communication between health care providers and patients, medication management, and informed consent. Health literacy means being empowered with knowledge about prevention, present health state, worst and best case scenarios with diagnoses, treatment options, and prognosis or outcomes. For patients, health literacy means being empowered with knowledge about prevention, present health state, worst and best case scenarios with diagnoses, treatment options, and prognosis or outcomes. For all health care workers, and Chinese medicine physicians and practitioners in particular, it means an increase in education, skills, knowledge and clinical experience in cardiology.

Public Health Theories and Models

(a) Communication Theories (see Table 8.1)

TABLE 8.1

Communication Theory	Definition
Organizational communication	This theory includes the structure of communication within and between agencies and businesses.
Social communication	This theory includes the method that organizations use to transmit health agendas.
Social marketing	This theory carries the agenda to encourage public thinking, attitudes and behaviors.
Media advocacy	This theory considers objectivity and neutrality in the spreading of public information.
Medical journalism	This theory considers accurate medical information, which influences governments, organizations and physicians.
Edutainment	This theory is a method of transmitting information to the public, as a form of education and entertainment.
Interpersonal communication	This theory includes the many methods of communication between people, and organizational structures that depend upon one another.

(b) Social Theories and Models (see Table 8.2)

TABLE 8.2

Public Health Research Theories	Definition
Diffusion of Innovations theory	This theory explains how information, opinion and behaviors are circulated amongst a population.
Health belief model	This theory explains how a belief influences patterns of behavior, whether the belief is right or wrong.
Social cognitive theory	This theory explains how having personal efficacy, self motivation and monitoring leads to success.
Theory of reasoned action	This theory explains how attitudes are predictors of behaviors.
Ideation	This theory explains how ideas are spread through peer approval and other factors that affect decision making.
Convergence theory	This theory explains the recognition that all industrialized countries tend to be more alike, which diminishes the possibilities of conflicts between them.
Stages of behavior change model	This theory explains the process that a hesitant target in a social system resorts to acceptance.
Communication for persuasive theory	This theory explains 12 steps for assimilating a particular behavior into a person's everyday life.

(c) Models for Strategic Behavior and Social Change Communications

Communication for behavioral impact explains how encouraging behavioral change in a society can save lives, i.e. stopping the spread of communicable diseases, and affecting risk factors for cardiovascular disease, lifestyle modification, etc.

The Precede–Proceed Model uses mnemonics to plan, design, implement and evaluate health programs (see Table 8.3).

Improving Patient Outcomes

In 1972 an article called *Doctor and Patient Communications* was written by Korsch and Negrete. Social science researchers began to develop theories which could encourage physicians to have more patient empathy.

(a) Social Support Theory

This theory explains the importance of enacted support given by health practitioners. Effective enacted support assists patients in adapting to stress and solving problems.

TABLE 8.3	
P = predisposing	P = policy
R = reinforcing and	R = regulatory and
E = enabling	O = organizational
C = constructs in	C = constructs in
E = educational	E = educational and
D = diagnosis and	E = environmental
E = evaluation	D = development

(b) Life Span Theory

This theory explains that a well-adjusted patient exhibits normal personality and relationships with others as a link to health state. Social relationships develop throughout life, and they include relationships with parents and others. The personality should be adaptive, with characteristics of low neuroticism and hostility, high optimism, coping and social skills. These skills are thought to be important for a healthy doctor support perception.

(c) Relational Regulation Theory

This theory explains the connection between the perception of poor health care provider support and bedside manner and mental health. Mental health is connected to regular social interactions.

See Table 8.4.

INTEGRATIVE MEDICINE PRACTICE

Cardiovascular disease is a major cause of serious health problems and death in the United States and worldwide. Western medicine practitioners have dedicated decades to developing examination and diagnostic methods, clinical protocols, treatment procedures and medicines. Despite these advances, the public is divided about the health outcomes. Some patients benefit while others who don't will require other options in addition to Western medicine care.

Traditional Chinese medicine (TCM) is a term used to describe theories, practices and treatments in Chinese medicine. Oriental medicine and Asian medicine are terms used for the traditional medical systems practiced throughout Asia. These systems share and also have their own

TABLE 8.4

Types of Support	Definition
Enacted support	This type of support takes a dominate role over the patient's wellbeing, and is provided for the children and the critically ill.
Perceived support	This support type represents the patient perception of the level and quality of care during examination and treatment. The patient will expect, empathy and sympathy from the health practitioner.
Empathizing	This support type involves understanding the patient's perspective. Sympathizing means demonstrating caring action.

unique acupuncture theories, practices, herbal pharmacopeia and treatment methods.

TCM is practiced and researched by many prestigious private and public institutions in the United States and worldwide. As a result, traditional Chinese medicine has become an important system of medicine. In the United States, Chinese medicine has been emerging since the 1970s as a recognized and licensed system of medicine. Practitioners must complete a rigorous 4-year master's degree program in Oriental medicine, which includes Chinese medicine, Korean and Japanese methods, as well as Western medicine. Some universities include homeopathy, Ayurvedic, Tibetan, etc. Learners must pass national board exams before applying for a state license to practice. Many prestigious universities provide postgraduate options to study further for their PhD in Oriental medicine, specializing in many areas such as sports medicine, gynecology and oncology. Practice options range from private practices to collaborations with Western medicine practitioners, forming integrative referrals or group practices.

Research and development of the integration of traditional Chinese medicine with Western medicine started in China in 1978. The *China Journal of Integration of TCM and Modern Medicine* started in China in 1981. Since then there have been many universities, research programs and academic organizations in China devoted to the development and advancement of integrative Chinese medicine in various areas of cardiology. These programs encourage study-abroad efforts to improve clinical knowledge and practice skills. In the United States, Oriental medicine practitioners, across many states, have varying degrees of practice scope which allow the treatment of patients using acupuncture, herbs, injection therapy, tui na, and other modalities. Specialty areas have also been established and have advanced treatment options in women's health, fertility, cancer treatment, stroke treatment and sports medicine. Efforts in research, health literacy and practitioner education are now necessary to shape attitudes and behaviors about Chinese medicine for cardiovascular care.

RESEARCH METHODS IN THE PRIVATE CLINIC SETTING

In 1965, Austin Bradford Hill had studied much about how one determines 'cause'; he then created criteria to determine this (Table 8.5). Clinical epidemiology is used to evaluate patient care practices and to help develop research studies (Tables 8.6 and 8.7); it addresses understanding about

TABLE 8.5

Criterion	Definition
Strength	The larger the association, the more likely it is causal.
Consistency	The more consistent the findings over different sample factors, the more likely it is causal.
Specificity	The more specific an association, the larger the probability of cause.
Temporality	The effect must occur after the cause.
Biological gradient	The wider the exposure the more prominent the effect; inversely the wider the exposure the lower the incidence.
Plausibility	The more complete the knowledge of the mechanism, the more understanding of the plausibility of an effect.
Coherence	Some public findings that can be created in controlled settings.
Experiment	Evidence is proven through controlled study.
Analogy	Different studies may have similar effects.

TABLE 8.6

Research Studies	Use
Observational studies	Observing disease patterns and frequency to evaluate association between risk factors and outcomes.
Interventional studies	Experiments and clinical trials which are conducted to explore the association between treatment interventions and outcomes.

disease incidence and prevalence, causes of diseases in communities, types of patient in the population, and the best methods to help them.

Health practitioners with private clinics devoted to treating patients with cardiovascular disease should endeavor to understand more about them. Different types of studies (Table 8.8) can guide the recruiting of new patients to study or in organizing an existing clinical caseload into study groups.

TABLE 8.7

Types of Research	Definition	Goal
Basic research	Explains the phenomenon being investigated.	To understand and explain a specific characteristic.
Applied research	Applies a scientific explanation to real-world situations.	To understand problems and potential solutions.
Summative research	Summates the cause and effect specifics of a study.	Summates the specific conditions under which an intervention was directly effective for a specific problem.
Formative evaluation	Evaluates the effectiveness of programs developing in their settings.	Evaluates the effectiveness of a program within the setting of its development.
Action research	Studying specific problems which are active in a location.	Studying specific problems within a program, setting or organization.

TABLE 8.8

Types of Study	Use
Cohort studies	Comparing patients with a risk factor or other exposure to others without them.
Case control studies	Comparing a case sample group with a disease with a case control sample group without the disease. Later a retrospective study analyzes what potential exposures that both groups encountered over time.
Case series	A study on the experience of a patient, or small group of patients with a similar diagnosis; or comparing periods during which they are exposed to some factor which produce an illness with periods when they are unexposed.

SUMMARY IN THE CONTEXT OF PERSONALIZING MEDICINE

Health communications are used to shape the knowledge and skills of the public in order to encourage preferred behaviors. Chinese medicine practitioners providing cardiovascular care should be prepared to use theories and models to increase favorable patient outcomes and help reduce the incidence, prevalence, and risk of cardiovascular disease.

FURTHER READING

[1] Wood J. Relational communication. 2nd ed. Belmont, CA: Wadsworth; 2000.
[2] Bryant CA, et al. Community-based prevention marketing: the next steps in disseminating behavior change. Am J Health Behav 2000;24:61–8.
[3] Hornik RC. Public health communication:evidence for behavior change. Mahwah, NJ: Erlbaum; 2002.
[4] Xilong P, et al. Service utilization in community health centers in China: a comparison analysis with local hospitals. BMC. Health Serv Res 2006;6:93.
[5] Krabbe PFM, et al. Are patients' judgments of health status really different from the general population? Health Qual Life Outcomes 2011;9:31.
[6] Fung ICH. Chinese journals: a guide for epidemiologists. Emerg Themes Epidemiol 2008;5:20.
[7] Thorpe C. Science against modernism: the relevance of the social theory of Michael Polanyi. Br J Sociol 2001;52:19–35.
[8] Earp JA, Ennett S. Conceptual models for health education research and practice. Health Educ Res 1991;6(2):163–71.

PART II

Factors in Prevention and Treatment

Chapter 9

Lipoprotein Disorders

Chapter Objectives

After studying this chapter you should be able to:

1. Explain the various related categories in Chinese medicine
2. Describe how hepatic, intestinal and renal function are primary mechanisms
3. Describe the secondary mechanisms
4. Discuss treatment methods in Chinese medicine
5. Summarize the context of the patient role in personalizing medicine

OVERVIEW

This chapter describes the effects and treatment of lipoprotein disorders within integrative cardiovascular Chinese medicine. Chinese medicine relies on Western medicine for the diagnostic methods and the direction for treating lipoprotein disorders. Hepatic, intestinal and renal disorders are well known primary factors related to lipid transport and accumulation. However, to improve outcomes, health care providers should consider secondary causes such as metabolic and hormonal disorders and pharmaceutical medications. These determine the current state of the body based on genetics, diet and lifestyle.

Metabolic and Hormonal Disorders

Hormonal causes of lipoprotein elevation include hypothyroidism, elevated estrogen levels in pre-, peri- and post-menopausal women, and sex hormones in men. The metabolic causes include the symptoms found in metabolic syndrome and type II diabetes, both of which tend to elevate cholesterol levels. Symptoms include elevated blood-pressure levels and LDL levels with reduced HDL levels.

Integrative Cardiovascular Chinese Medicine. DOI: http://dx.doi.org/10.1016/B978-0-12-420014-2.00006-5

Pharmaceutical Medications

Certain medications – corticosteroids and anabolic steroids, beta blockers, thiazide diuretics, estrogen, immunosuppressive and antiretroviral drugs, and certain antipsychotic medications – cause an increase in triglyceride levels and lower HDL levels.

The primary clinical presentation which is indicated is the average total cholesterol with high triglycerides, average LDL (low-density lipoprotein) and low HDL (high-density lipoprotein). Many genetic determinants, including familial hypercholesterolemia (FH), have been proven to be responsible for obesity, cardiovascular disease and other diseases which complicate treatment.

The main concern is to protect and monitor the arteries against atherosclerosis and coronary artery disease.

Key Points

- Hyperlipidemia is considered a pathologic deficiency between the liver, spleen and kidneys.
- On cell membranes phospholipids are components of messenger sites.
- Genetic determinants predispose for accumulation of triglycerides, LDL and cholesterol at specific sites in the body, which contribute to renal, hepatic and cardiovascular diseases.
- The liver and intestinal pathways are part of a complicated system for the transport of lipoproteins and cholesterol, to specific sites of uptake, for the production of hormones and bile.
- Patients with chronic renal failure often have elevated triglyceride levels with a low HDL, while those at end stages on dialysis often present with signs of atherosclerosis and coronary artery disease.
- Macrophages attack *Staphylococcus aureus* at infection sites. The macrophage foam cells leave cholesterol streaks in the arteries and cause endothelial damage.
- There are five types of genetic disorder which affect lipid production and are associated with a predisposition to cardiovascular and other diseases:
  - Type I is hyperlipidemia. It is a genetic disorder marked by increased transport of cholesterol on the chylomicrons produced in the small intestine and pancreatitis.
  - In type II, some patients may develop type II diabetes mellitus, hypothyroidism, atherosclerosis and coronary artery disease.
  - Type III involves a genetic mutation of the *ApoE* gene and is evidenced by the presence of various xanthomas.
  - In type IV, cholesterol may be stable while triglycerides, including LDL, are elevated, and it may involve FH.
  - Type V is a combination of types I and IV.

- A defect in the proprotein convertase subtilisin kexin type 9 (PCSK9) gene leads to a reduction in cholesterol in cells. This gene may be more prevalent in the African American population.
- Elevated LDL levels complicate the treatment of diabetes mellitus.

PATHOGENESIS

Chinese Medicine

In Chinese medicine lipoprotein disorders are categorized as tanzhuo (tan-joo-oh), xueyu (shoo-eh-yoo) and xuan yun (shoo-an-yoo-n), and are classified under blood stasis and turbid phlegm obstruction. Because it is a factor in other diseases, it may also belong to the categories of headache, stroke, palpitation, obstruction and vertigo. Like hypertension, hyperlipidemia is more a symptom than a disease itself. Hyperlipidemia is considered a pathologic deficiency between the liver, spleen and kidneys. With depletion of liver yin, hyperactive liver yang overacts upon the spleen – resulting in an acute digestive impairment. Prolonged excess liver yang overaction results in chronic digestive problems.

Western Medicine

The lipid transport system is involved in the movement of lipids from the hepatobiliary and gastrointestinal systems to cellular uptake sites where hormones are produced. Triglycerides are needed for energy. Cell membranes utilize phospholipids as components of messenger sites for the regulation of mechanisms functioning on the surface. Cell membranes, bile and hormones all need cholesterol. When genetic, diet and lifestyle factors cause lipoprotein disorders, the result is hyperlipidemia. There are five types of hyperlipidemia: types I–V; they are associated with hepatosplenomegaly, xanthomas, carbohydrate metabolism problems and obesity.

Type I is a genetic disorder marked by (i) increased transport of cholesterol on the chylomicrons produced in the small intestine, and (ii) pancreatitis. In type IV, cholesterol may be stable while triglycerides, including LDL, are elevated and may involve familial hypercholesterolemia (FH). Type V is a combination of types I and IV.

Type II

This type involves defects in LDL receptors which control inter- and extracellular reception of cholesterol, causing an elevation in blood plasma. Some patients may develop type II diabetes, hypothyroidism, atherosclerosis and coronary artery disease through FH; for others it may develop through lifestyle and diet.

Type III

This type is hyperlipoproteinemia. It involves a genetic mutation of the *ApoE* gene and increased production of LDL, leading to facial and palmar xanthomas.

DETERMINANTS AND MECHANISMS

Genetic Determinants

1. Let-7. This is a class of miRNA which accumulates in fat tissues and predisposes individuals to insulin resistance, obesity and type II diabetes mellitus.
2. Familial hypercholesterolemia (FH). These genetic determinants include a predisposition for accumulation of triglycerides, LDL and cholesterol at specific sites on the body. They contribute to renal, hepatic and cardiovascular diseases.
    - (a) Familial hyperchylomicronemia (type 1). This genetic determinant involves lipoprotein lipase (LPL) deficiencies leading to elevations in triglycerides and reduced HDL levels.
    - (b) Familial hypercholesterolemia (type II). This genetic determinant includes various genetic defects. A defect in the proprotein convertase subtilisin kexin type 9 (PCSK9) gene leads to elevated cholesterol levels. This gene may be more prevalent in the African American population in the United States. A defect in the LDL-R receptor gene causes elevated LDL levels. Apo B gene defects lead to high elevations of LDL. Lipoprotein A Lp(a) includes Apo B, and an abundance in plasma is linked to coronary artery disease.
    - (c) Familial hyperlipoproteinemia (type III). This genetic determinant includes elevated LDL levels and is identified through the *Apo E2 genotype; it is linked to coronary artery disease.
    - (d) Familial hyperlipoproteinemia (type IV). This genetic determinant includes Apo B (which binds to LDL receptors), PCSK9 mutations, LDLRAP1 mutations and ABCG 8 mutations. Overproduction of VLDL by the liver leads to elevations in cholesterol levels. Symptoms present as an accumulation of xanthomas, higher cholesterol levels and metabolic syndrome.
    - (e) Type V may be a combination of type I and IV. Symptoms include splenomegaly and xanthomas.
3. Apolipoprotein B (Apo B) isoforms B100 and B48.
    - (a) Apo B100: found in the liver (VLDL and LDL) and high levels are associated with coronary artery disease due to macrophages responding to *Staphylococcus aureus* infection. Macrophage foam cells leave cholesterol streaks in the arteries and cause endothelial damage.
    - (b) Apo B48: found in chylomicrons which transport lipoproteins from the small intestine to the body through the blood flow. Elevated Apo B48 raises LDL levels and complicates diabetes mellitus treatment.

TABLE 9.1

Constitution	General Incidence of Lipoprotein Disorders
Teenage and young adult	Childhood obesity, Type I Diabetes, insulin resistance and hormonal disorders.
Middle-age	Development of Type II Diabetes, hypothyrodism and other sex-hormonal disorders, obesity, hypertension, insulin resistance, renal symptoms, gastrointestinal disorders.
General geriatric	Coronary artery disease, renal failure, hepatic failure, gastrointestinal diseases.

4. *Apolipoprotein E (Apo E) allele mutations E2, E3 and E4.
   (a) E2: slow fat metabolism, increased in risk of coronary artery disease, Parkinson disease, coronary artery disease.
   (b) E3: neutral allele often subjectively paired with E4 for clinical significance.
   (c) E4: various symptoms, including atherosclerosis, sleep apnea, ischemic cerebrovascular disease, multiple sclerosis, Alzheimer's disease.

Constitutional Factors

Issues within the constitution involved genetics, sex, race, age and lifestyle.

The general incidence of lipoprotein disorders is shown in Table 9.1.

Mechanisms

(a) Hepatic and intestinal pathways. The liver and intestinal pathways are part of a complicated system for the transport of lipoproteins and cholesterol to specific sites of uptake for the production of hormones and bile. After food has been processed in the stomach and moved towards the jejunum, bile from the hepatic system is excreted. HMG-CoA reductase helps to produce cholesterol in the liver. Cholesterol (VLDL) is sent from the liver into the small intestine where it is processed into chylomicrons, and deposited in the portal vein. The chylomicrons undergo a complicated lipoprotein exchange process related to the reverse cholesterol transport pathway, so that the liver can later metabolize fats.

(b) Renal and hepatic disorders. Patients with chronic renal failure often have elevated triglyceride levels with a low HDL, while those at end stages on dialysis often present with signs of atherosclerosis and coronary artery disease. Patients with glomerulonephritis may have problems with liver lipid metabolism and bile secretion, causing the elevation of

LDL level. In liver failure, cirrhosis may result in lipoprotein X, which is found in LCAT deficiency. Symptoms may present on the face around the eyelids as xanthomas.

EXAMINATION OF PATIENT

Visual Observation

Look for areas of edema and fat accumulation, including the face and body, especially the waistline.

Observation

- Head: dizziness, distended feeling, heavy feeling, brain fog and forgetfulness
- Eyes: xanthomas
- Mouth: bitter taste, sticky taste sensation, dry throat, greasy tongue coating
- Emotions: vexation, moody and temperamental restlessness
- Limbs: pain, numbness and tingling
- Others: hormonal problems, weight gain

Physical Examination

Measure the waist circumference and weight. BMI can be used if it is a significant part of regular practice.

Blood Pressure

Evaluate the effects of blood pressure by viewing the retina, which can predict future patterns of cardiovascular disease. This is taken in sitting, standing or supine positions, and readings indicate possible hypertension.

Laboratory Tests (Tables 9.2–9.4)

WESTERN MEDICINE

General Hyperlipidemia

(a) Clinical presentation. High-fat diet, overweight or obesity, heavy consumption of alcohol, and an inactive lifestyle with lack of exercise.

(b) Work up and treatment. This patient with rising triglyceride and VLDL levels has a risk factor for other diseases such as certain thyroid problems, ovarian cysts, elevated female hormone levels and kidney disease.

This type of patient is usually monitored with regular suggestions for lifestyle and diet modifications.

TABLE 9.2

Lipid type	mg/dL	mmol/L	Interpretation
Total cholesterol	<200	<11.1	Normal
	200–239	11.1–13.3	Borderline
	>240	>13.3	High
LDL	<100	<5.5	Normal
	100–129	5.55–7.15	Good
	130–159	7.15–8–82	Borderline
	160–189	8.82–10.5	High
	>190	>10.5	High risk
HDL	<40	<2.21	High risk
	41–59	2.21–3.27	Borderline
	60	3.27	Good

TABLE 9.3

Type	Laboratory Findings
Familial hyperchylomicronemia (Type I)	Increased total cholesterol
	Normal or increased LDL
	Normal or decreased HDL
Familial hypercholesterolemia (Type II)	*Subtype a:* increased LDL
	Subtype b: increased LDL and total cholesterol
Familial hyperlipoproteinemia (Type III)	Increased total cholesterol
	Normal or decreased LDL
Familial hyperlipoproteinemia (Type IV)	Increased total cholesterol
	Increased LDL

- Increase physical activity: continuous movement 30 minutes per day
- Diet: use the dietary approaches which increase fruits, vegetables, grains and reduce meat, sugar and dairy products
- Weight maintenance: target as close to ideal BMI as possible

TABLE 9.4

Lab	Detections
C-reactive protein serum	Strong predictor of cardiovascular disease. LDL reduction also alters/ adjusts the prediction.
Triglycerides	Early indicator of hyperlipidemia and coronary artery disease.
Thyroid panel	Tests TSH/ T4 levels.
Kidney function tests: Urinary albumin Plasma rennin BUN	Risk factor or monitoring for nephropathy diabetic nephropathy Hypokalemia and acute or chronic kidney disease.

- Stop smoking: even though smokers feel that it relaxes, nicotine increases heart rate
- Reduce or avoid alcohol: drink in moderation or avoid drinking altogether

Medications

- Ezetimibe
- Statins
- Bile acid sequestrants
- Niacin
- Fibrate

Familial Hyperchylomicronemia (Type I)

(a) Clinical presentation. alcoholism, xanthomas, abdominal pain, lipemia retinalis.
(b) Treatment:
  - Ezetimibe
  - Statins

Familial Hypercholesterolemia (Type II)

(a) Clinical presentation:
  - Subtype a: hypothyroidism, xanthomas
  - Subtype b: obesity, coronary artery disease, types I and II diabetes mellitus, nephritic syndrome

(b) Treatment:
- Type a: bile acid sequestrants, statins and niacin
- Type b: statins, niacin, fibrate

Familial Hyperlipoproteinemia (Type III)

(a) Clinical presentation: hypothyroidism, uncontrolled diabetes mellitus and nephritic syndrome, alcoholism, xanthomas, abdominal pain, lipemia retinalis

(b) Treatment:
- Ezetimibe
- Statins

Familial Hyperlipoproteinemia (Type IV)

Dietary control.

(a) Clinical presentation: uncontrolled diabetes mellitus, alcoholism, estrogen therapy, hyperthyroidism, alcoholism, xanthomas, abdominal pain, lipemia retinalis

(b) Treatment:
- Niacin
- Statins

CHINESE MEDICINE

Tan zhuo zhang zu and gan shen kui types

See Figure 9.1.

(a) Clinical presentation: blurred vision, irritability, dizziness, soreness and weakness of the limbs, low back, knees and genitalia, vertigo, tinnitus, insomnia, forgetfulness, dream-filled sleep, poor appetite

Tongue = scanty fur

Pulse = thready and tight

See Figure 9.2.

Deficiency of Liver and Kidney Yin

Deficient kidney yin fails to nourish the liver yin. Liver yin deficiency as well as emotional and psychological issues can quickly cause liver yang hyperactivity and qi stagnation. The liver overacts upon the stomach leading to digestive problems.

FIGURE 9.1

Clinical Pearl:

The liver and kidney need to be as balanced as possible to allow for smooth blood flow and natural elimination of excess lipids through the digestive tract.

FIGURE 9.2

Dampness retention leading to damp-heat

Spleen and kidney yang deficiency leads to excess fluid retention in the body. Fluid which is not eliminated through the kidneys or the intestines lingers in the interstitium, insulates and generates precipitation. The resulting moist heat may cause lin syndrome imitating a sensation of infection or inflammation.

FIGURE 9.3

Clinical Pearl:

On observation, the patient may have obesity or edema. The lower body proportion may appear larger than upper body.

FIGURE 9.4

(b) Treatment:
- Therapeutic principle: nourish yin, subdue yang
- Medicine: Fleece flower formula 3 × daily between meals with warm water
  - Weicao tang 2 × daily between meals with warm water
- Exercise: Tai Qi, qigong and stretching 2 × daily for 15 minutes

(c) Medicated diet:
- Avoid accompanying greasy food with herbal medications
- Oat or rice gruel which includes hawthorn or kelp daily during breakfast

Gan yu pi xu type

See Figure 9.3.

(a) Clinical presentation: irritability or depression, headache or head distention, bitter and dry mouth, poor appetite, heat sensation in the chest, heavy body feeling, lower abdominal and low-back pain with false sensation of urinary urgency with dark scanty urine and dry feces.

Tongue = pale with yellow greasy coat
Pulse = rolling or smooth

See Figure 9.4.

(b) Treatment:

- Therapeutic principle: eliminate pathogenic dampness and drain the heat
- Medicine: Gentian Formula 3 × daily between meals with warm water
  - Eliminate Blood Lipid Formula 3 × daily between meals with warm water
  - Yinxing tang 2 × daily with warm water Shen qi wan 3 × daily with warm water
  - Gao lin tang 3 × daily with warm water
- Exercise: Tai Qi, Qigong or stretching 2 × daily 15 minutes

(c) Medicated diet:

- Avoid greasy food with these herbal medications
- Oat or rice gruel which includes hawthorn, lotus leaf or plantain tuber daily during breakfast

See Figure 9.5.

(a) Clinical presentation: obesity, dizziness, chest and abdominal fullness, sensation of heavy head and body, phlegm accumulation in the sinuses prompting drawing and spitting, tinnitus, a bitter taste in the mouth with a sticky feeling in the mouth with need to drink more water, and poor appetite.

Tongue = pale purple with ecchymosis
Pulse = wiry or rolling

See Figure 9.6.

(b) Treatment:

- Therapeutic principle: break up phlegm, resolve stasis
  - Medicine: with greasy food

Phlegm Stagnation and Stasis

Spleen and kidney yang deficiency leads to excess fluid retention. Prolonged dampness retention leads to phlegm and fat production in the body.

FIGURE 9.5

Clinical Pearl:

This patient may have oily sweat which may also suggest a bacterial infection and possible inflammation. Include antibacterial and anti- inflammation herbs in the medicine.

FIGURE 9.6

- Oat or rice Reduce Blood Lipid Formula 3 × daily between meals with warm water
  - Ginko leaf formula 2 × daily with warm water
- Exercise: Tai Qi and Qigong 2 × daily 15 minutes

(c) Medicated diet:
- Avoid greasy foods during meals
- Gruel which includes hawthorn, kelp or other medicinal herbs for daily breakfast

MEDICINES

A. Western medicine drugs

See Table 9.5.

B. Chinese medicine formulas

These formulas can be modified with herbs in the pathologic listings (Chapter 19). Modifications allow for practitioners to adjust a formula according to clinical presentations while maintaining a course of treatment.

See Table 9.6.

SUMMARY IN THE CONTEXT OF PATIENT PERSPECTIVES FOR PERSONALIZING MEDICINE

For integrative medicine to provide effective personalized medical solutions for patients with high-risk lipoprotein disorders, the course of treatment may need to include either Western medicine or Chinese medicine – but not both at the same time. Patients with what is considered to be a general weight problem (due mostly to lifestyle and diet) or who may need to lose 20 lbs (40.4 kg) may see quicker, more favorable benefits within this treatment.

Patients with cardiovascular disease and familial hypercholesterolemia may experience some challenges during treatment, especially with renal, hepatobiliary disorders and diabetes mellitus.

TABLE 9.5

Medicine	Action
Ezetimibe	Lowers cholesterol absorption in the small intestine.
Statins	Inhibits HMG-CoA reductase enyme for cholesterol reduction.
Bile acid sequestrants	Prevents fat absorption in the intestines.
Niacin	Helps breakdown fat tissue.
Fibrate	Combines with statins to treat high cholesterol.

TABLE 9.6

Medicine	Action
Fleece flower formula	Reduces liver hyperactivity, relieves body distention, nourishes kidney yin and blood.
Weicao tang	Reduces blood lipids and increases lipid metabolism.
Gentian formula	Purges liver fire to reduce adverse effects on the heart, pathogenic dampness transforming into heat.
Eliminate blood lipid formula	Reduces blood lipids for weight loss.
Shen qi wan	Helps to tonify and strengthen the kidneys.
Reduce blood lipid formula	Reduces LDL levels.
Ginko leaf formula	Assists with the symptoms of lipoproteinemia, fights free radicals, moves blood and breaks up phlegm in the vessels.

Chinese medicine practitioners can take several pathways to guide treatment. Two general ones include either of the following two pathways. (1) Putting the patient on a treatment course for a month and evaluating the results – then modifying the treatment for an additional month. This may work well for a patient who is in the early stages of cardiovascular disease and who likely does not have other diseases which may complicate the course. (2) Regular weekly visits with differentiating and modifying formulas where necessary. This may work better for patients with mild to moderate renal, hepatobiliary disorders, diabetes mellitus, hypertension, etc. who may be under the care of a Western medicine practitioner where regular monitoring is flexible.

A treatment course using Chinese medicine alone may necessitate a close regular assessment of laboratory blood results. In addition, addressing hypothyroidism by providing TCM treatment for reducing TSH levels may assist in diagnosis, the steady harmonizing of lipoproteins, and aiding in loss of excess fat. In addition, genetic mutations which predispose certain patients to lipoprotein disorders and associated systemic diseases need to be addressed. For example, remember that the APO E gene is found on chromosome 19 with three allele mutations E2, E3 and E4. Consider whether the patient's race, in relation to disease presentation, may give a clue if stubborn symptoms persist. Regular blood work provides insight into the progress of treatment – which supplements tongue and pulse diagnosis. Noticing the evidence that your treatment is giving good results helps the patient to cooperate in organizing a dietary and lifestyle program, which can further personalize the treatment plan.

FURTHER READING

[1] Hu J, Yang Z, Ren K, et al. Clinical study on hyperlipemia treated by weicao capsule. Mod J Integr Tradit Chin West Med 2010;2010–25.

[2] Bonow RO, Mann DL, Zipes DP, Libby P, editors. Braunwald's heart disease: a textbook of cardiovascular medicine. 9th ed. Philadelphia, PA, USA: Saunders Elsevier; 2011.

[3] Daniels SR, Greer FR. Committee on nutrition. Lipid screening and cardiovascular health in childhood. Pediatrics 2008;122:198–208.

[4] Cohen J, Pertsemlidis A, Kotowski IK, et al. Low LDL cholesterol in individuals of African descent resulting from frequent nonsense mutations in PCSK9. Nat Genet 2005;37:161–5.

[5] Abifadel M, Varret M, Rabès JP, et al. Mutations in PCSK9 cause autosomal dominant hypercholesterolemia. Nat Genet 2003;34:154–6.

[6] Victor RG, Haley RW, Willett DL, et al. The Dallas Heart Study: a population-based probability sample for the multidisciplinary study of ethnic differences in cardiovascular health. Am J Cardiol 2004;93:1473–80.

[7] Chang AY, Abdullah SM, Jain T, et al. Associations among androgens, estrogens, and natriuretic peptides in young women: observations from the Dallas Heart Study. J Am Coll Cardiol 2007;49:109–16.

[8] Dubuc G, Chamberland A, Wassef H, et al. Statins upregulate PCSK9, the gene encoding the proprotein convertase neural apoptosis-regulated convertase-1 implicated in familial hypercholesterolemia. Arterioscler Thromb Vasc Biol 2004;24:1454–9.

[9] Ridker PM, Hennekens CH, Buring JE, et al. C-reactive protein and other markers of inflammation in the prediction of cardiovascular disease in women. N Engl J Med 2000;342:836–43.

[10] Ridker PM. High-sensitivity C-reactive protein: potential adjunct for global risk assessment in the primary prevention of cardiovascular disease. Circulation 2001;103:1813–8.

[11] Cushman M, Legault C, Barrett-Connor E, et al. Effect of postmenopausal hormones on inflammation-sensitive proteins: the Postmenopausal Estrogen/Progestin Interventions (PEPI) study. Circulation 1999;100:717–22.

Chapter 10

Hypertension

Chapter Objectives

After studying this chapter you should be able to:

1. Explain how the liver, heart and kidneys are involved in hypertension in both Chinese medicine and Western medicine
2. Describe how genetic determinants involve salt and water excretion
3. Discuss the vascular, renal and neural mechanisms
4. Discuss suggested Chinese medicines which can be modified for personalizing medicine
5. Summarize the context of personalizing medicine

OVERVIEW

This chapter outlines the aspects of integrative medicine which may be important in developing personalized approaches for patients with hypertension. Hypertension is considered when blood pressure is >140/> 90 in a seated or resting individual over more than one visit to the health care practitioner. There are two types of hypertension including essential or idiopathic. Generally, lifestyle choices and environmental factors and other underlying diseases are contributors. In secondary hypertension the causes are identified to be directly related to endocrine pathology, renal disease, coarctation of the aorta, sleep apnea, and adrenal insufficiency.

In Chinese medicine, hypertension is considered to be a direct injury to the heart, liver and kidneys. It is a cyclic condition marked initially by imbalances between yin and yang due to prolonged emotional crisis, improper diet, and other factors resulting in the deficiency of both. Therefore, severity of incidence is often subjective to the overall condition of the individual. Hypertension is a serious risk factor in cardiovascular disease, and yet it is considered to be highly treatable. According to the Joint National Committee on Prevention (JNC7), Detection, Evaluation, and Treatment of High Blood Pressure, in adults, the acceptable systolic pressure should be <120 and the

Integrative Cardiovascular Chinese Medicine. DOI: http://dx.doi.org/10.1016/B978-0-12-420014-2.00007-7

diastolic should be <80. Stage 1 hypertension is considered to be between 140 and 160 systolic and between 90 and 100 diastolic. Stage 2 is considered to be between 160 and 180 systolic and between 100 and 110 diastolic. Crisis is considered to be >/ = 180/110.

Lifestyle, gender and age may have an influence in adults, with a common tendency of systolic pressure to rise over time. In elderly individuals, arterial morbidity increases, causing a rise in blood pressure. According to both Chinese and Western medicine, blood pressure is affected by psycho-emotional responses, diet and digestion, lifestyle choices and exercise. In addition, the time of day, season of the year, environmental pollution and microbes are also contributing factors.

Left untreated or under-treated, hypertension opens the risk for other cardiovascular diseases, renal failure, brain and neurologic diseases, vision problems and other organ failures.

In Chinese medicine, pathologic factors, severe emotions, improper nutrition and visceral yin and yang disharmonies can explain some of the reasons for complications in controlling hypertension. In Western medicine, renal disorders, hormonal disorders (e.g. aldosterone and estrogen deficiencies), are other factors which will require closer evaluation in patient management.

Key Points

- A genetic predisposition or family history of hypertension increases risk.
- Prolonged stress and anxiety related to environment can also increases risk.
- Pre-hypertension is an early warning to modify lifestyle.
- Pre-existing conditions, such as thyroid disease and sleep apnea, may be a cause, and can be managed with modifications in lifestyle.
- In Chinese medicine, hypertension is included in the categories of xuan yuan <shoo-an> (vertigo) and tou tong <toe-tung> (headache).
- Certain Western and Chinese medications for asthma, such as bronchodilators and ma huang (ephedra), cardiotonics, formula harmonizers such as gan cao, and antioxidants such as 100% lu cha (green tea) – when used without professional supervision – can cause temporary elevations in blood pressure.
- Lifestyle contributors: lack or insufficient physical activity, excessive smoking, excessive alcohol consumption, excessive sodium intake, insufficient balancing nutrients and other minerals.
- Use of the birth-control pill and hormone replacement.
- Current history of obesity.
- In the USA, the occurrence of hypertension remains higher in adult males than in females at any age; and in African Americans, Hispanics and Asians than in Caucasians.

- In China, all Han men as well as the other 56 ethnic minorities of men, are five times as likely to develop and die of hypertension-related complications compared to women.
- Treatment of hypertension requires a close physician-and-patient team relationship with high patient interest and compliance.
- The JNC7 classification refers to an adult who is not considered acute or within an emergency hypertensive status, and untreated with medication or dietary restrictions.
- From the normal range, each increment of 20 mmHg/10 mmHg increases cardiovascular disease risk two and half times.
- Patients with cardiovascular, diabetes or renal diseases are under different criteria: the target should be $<130/<80$ mmHg.
- Systolic elevation, rather than diastolic, is a stronger risk factor in individuals over 50.
- Two or more medications, in either Western medicine or Chinese medicine, or in integrated medicine, may be needed to control blood pressure in some patients. Chinese medicine is considered to be most effective up to stage 1. In late stage 2, or critical events $>160/100$, close monitoring and Western medicine drugs may be indicated – with diuretics especially in more serious disease complications.

PATHOGENESIS

Chinese Medicine

In Chinese medicine, hypertension is classified under vertigo or xuan yun and headache or tou tong and is more a symptom rather than a disease itself. Hypertension expresses an imbalance between yin and yang among the heart, liver and kidneys in stages. Severe emotions, improper nutrition, organ system disharmonies and metabolic imbalances contribute to determining the prognosis.

Yin and yang are interdependent and interchangeable. Both are the driving forces of qi energy and physical substance, and can explain the simplicity and complexity of body functioning. The heart, kidney and liver are yin organs with yin and yang responsibilities. They maintain the yang spirit, energy and vitality and the circulation of yin blood and fluids. Hypertensive feelings of apathy, listlessness, heaviness, tiredness, forgetfulness and amnesia belong to yin and are associated with heart syndromes. Restlessness, irritability, anger and vexation belong to yang and are associated with liver syndromes. Both are associated with kidney essence, yin and yang deficiencies. Deficiency of either yin or yang results in exaggeration in the characteristics of the other, interchangeably, until they are both completely consumed.

TABLE 10.1

JNC7 Prevention, Detection, Evaluation, and Treatment of High Blood Pressure		
Classification	**Systolic mmHg**	**Diastolic mmHg**
Normal	<120	<80
Pre-Hypertension	120–139	80–89
Stage 1 Hypertension	140–159	90–99
Stage 2 Hypertension	>160	>100

Western Medicine

See Table 10.1.

The JNC7 classification determines essential hypertension as a slight elevation in blood pressure (120–139/80–89) which then gradually rises over a period of years. The factors involved in essential hypertension include:

- Heredity: hypertension occurring in families through genetics, or lifestyle habits
- Environmental: this includes sedentary lifestyle, diet and nutritional deficiencies, occupational stress, smoking and alcohol consumption
- Demographic: age, race, gender, weight
- Health state: glucose levels, cholesterol levels, cancer, tumors, inflammatory conditions

Secondary hypertension develops suddenly as a result of underlying adverse health conditions or as a response to medication. Factors involved include:

- Renal complications: problems in processing sodium and fluid; problems with secretion of vaso-active materials; vasoconstriction due to angiotensin elevation
- Renin: excess secretion by gland tumors; suppressed or elevated plasma renin activity; coarctation of aorta
- Endocrine: aldosteronism, estrogen deficiencies
- Medications: certain prescription drugs, birth-control pills, respiratory medicine, certain cold and flu medicines and illegal drugs.

DETERMINANTS AND MECHANISMS

Genetic Markers

Hypertension is a major clinical concern across all ethnicities. Despite measures for treatment, there persists the issue of whether the control of hypertension is a preventative against heart failure. In some patients with persistent hypertension it may be necessary to consider the deficiencies in innate essence or genetics as they affect kidney functioning.

In Chinese medicine, ions involved in re-absorption have a yin and yang relationship. For example, chloride may be more yang, and potassium may be more yin. Mendelian forms of hypertension and hypotension are directly related to renal sodium re-absorption. The renin–angiotensin system and sodium-handling genes are associated with determining blood pressure.

MYH9

This gene is related to nephritic syndrome in which the blood pressure becomes a critical concern, leading to kidney failure. Moreover, natriuresis is a compensation mechanism marked by excessive excretion of sodium and water with increases in blood pressure. In Bartter and Gitelman syndromes, the patients require aggressive therapy with salt and potassium replacement. In Bartter syndrome, symptoms present in infants and children. In Gitelman syndrome, the symptoms present in children or later in life in adults. In Liddle syndrome, hypertension is autosomal dominant, with hypertension that often begins in childhood, featuring hypokalemia and aldosterone deficiency. In pseudohypoaldosteronism type 1, autosomal dominant and recessive forms are seen in newborn infants and in failure-to-thrive children.

STK39

This gene has recently been identified as a marker for essential hypertension in Amish patients.

Constitutional Factors

The personal habits of most patients are often directly related to hypertension outside of genetic factors. Inactivity and lack of exercise increases risk as it relates to an increase in body mass for around 50% of cardiovascular patients. The nicotine in cigarette smoke raises blood pressure by up to 20 mmHg.

See Table 10.2.

Mechanisms

(a) Vascular mechanisms. Vascular remodeling is a part of endothelial dysfunction. Vasoconstriction leads to increases in blood pressure. Arterial remodeling occurs due to the presence of collagen and vascular dilation.

(b) Renal mechanisms. The renin–angiotensin system is a hormonal mechanism which involves the activation of renin–angiotensin–aldosterone (RAAS) ACE, a product of this system which is found within the lungs, heart and all vessels, and which directly contributes to hypertension.

TABLE 10.2

Constitution	General Incidence of Hypertension
Teenage and young adult	Systolic hypertension and isolated systolic hypertension is the concern during this period. Overactive sympathetic nervous system responses in young men may foreshadow hypertension later in life. Aldosterone levels should be monitored in both young men and women for heart failure prevention.
General geriatric	Diastolic pressure may be dropping gradually during this time leading to hypotension in some patients. Vascular remodeling affects blood flow through the arteries. Isolated systolic hypertension is likely developed, especially in women, and can lead to heart failure.

Angiotensin II has two AT receptor types. AT1 is found in heart, brain, kidney, adrenal and liver and vessels and contributes to insulin resistance and vascular remodeling. AT2 is involved in aldosterone secretion and it contributes to the control of blood pressure. AT1 receptors express accumulated T-cells in adipose tissue and may assist with obesity-related hypertension. Prorenin is considered to be an inactive precursor to renin and does not bind to angiotensin. However, circulating prorenin binding to receptors in the heart and kidneys may increase TGF-β production, which contributes to vascular remodeling.

(c) Neural mechanisms. Neurogenic hypertension is caused by sleep apnea due to elevated catecholamine levels. Baroreflex failure is commonly seen in elderly patients, and it often occurs after meals. Sympathetic innervation disorders related to excessive activity are often seen in primary hypertension and diabetes type 2.

EXAMINATION OF THE PATIENT

Visual Observation

Perform a neurologic examination for stroke by noticing the patient's gait upon entering the room and bilateral grip strength.

Listening

- Assess memory, judgment and speech for neurologic problems before or after treatment with hypertensive drugs
- Head: distended feeling, insomnia day dreaming, brain fog and forgetfulness

- Eyes: vision problems, flushed eyes
- Ears: dizziness, tinnitus
- Mouth: bitter taste, dryness, tongue color is red or pale with a dry coating
- Emotions: vexation, moody and temperamental restlessness
- Limbs: pain, numbness and tingling
- Others: hormonal problems, climacteric events, sexual problems

Physical Examination

Measurement of temperature, waist circumference, weight and height.

Evaluate the effects of blood pressure by viewing the retina, which can predict future patterns of cardiovascular disease.

(a) Blood pressure:
  - This is taken in a sitting, standing or supine position, and the readings indicate stages of hypertension

(b) Pulse in TCM:
  - Chinese pulse diagnosis:
    - rapid = true heat, yin deficiency heat, infection
    - forceful = compensation due to deficiency
    - weak = deficiency
    - wiry = hyperactive liver condition
    - thready = hyperactive kidney

(c) Bruits indicate future vascular stenosis:
  - Neck bruit may indicate carotid stenosis
  - Abdominal bruit may indicate reno-vascular disease
  - Femoral bruit is a sign of atherosclerosis. Confirmation is given through taking the blood pressure in the lower leg

Auscultation

(a) Heart:
  - First palpate the apical impulse for left ventricular enlargement or hypertrophy
  - Listen for an early sign of hypertension, the fourth heart sound (S4), which indicates left atrium overwork
  - Listen for underlying sign of heart disease, the third heart sound (S3), which indicates left ventricular malfunction function

(b) Lungs:
  - Rales are unusual in hypertension, but their presence can also be heard, through (S3) gallop

Laboratory Tests

Routine Tests

12-lead electrocardiogram: in hypertension, the terminal portion of the P wave has a duration of 0.04 seconds and a depth of 1 mm.

See Table 10.3.

Other Tests

Echocardiography is used when left ventricular hypertrophy is suspected, and there is ambulatory hypertension and white-coat hyertension.

See Table 10.4.

TABLE 10.3

Lab	Detections
Urinalysis	Kidney function, protein, blood and glucose
Hematocrit	Anemia as found in inflammation and cardiovascular disease
Serum potassium	Hypokalemia as in aldosterone excess
Serum creatinine	Deficiency signal kidney disease
Blood glucose	Increased values signal diabetes
Lipid profile	High triglycerides, LDL, low HDL

TABLE 10.4

Lab	Detections
Serum uric acid	Risk factor or monitoring for gout
Urinary albumin	Diabetic nephropathy and hypertension
Plasma renin	Hypokalemia may signal aldosterone excess
Serum creatinine	Aldosterone excess, hypokalemia (when a diuretic was included in a failed drug regimen to reduce blood pressure)
Thyroid panel	Tests TSH/T4 levels
Serum calcium	Renal disease and monitoring of supplementation levels

PRE-HYPERTENSION

Western Medicine

(a) Clinical presentation: sleep apnea, dyslipidemia, kidney problems, increased emotional stress, endocrine imbalance, adrenal insufficiency.

Blood pressure: 120/80–139/89.

(b) Work up and treatment:

- Family history of hypertension, heart disease or other systemic diseases
- BMI >30
- Mild hyperlipidemia
- Age: women >45–55, men >40–60
- Smoker and alcohol drinker
- High stress relationships, lifestyle or occupation

This type of patient is usually monitored with a yearly physical examination with suggestions for lifestyle and diet modifications. *No medication is given.*

- Increase physical activity: continuous movement 30 minutes per day
- Diet: use the 'Dietary Approaches to Stop Hypertension' (DASH) diet
- Weight maintenance: target as close as possible to ideal BMI
- Stop smoking: although smokers feel that it relaxes, nicotine increases heart rate
- Reduce or avoid alcohol: drink in moderation or avoid drinking

See Figure 10.1.

Chinese Medicine

See Figure 10.2.

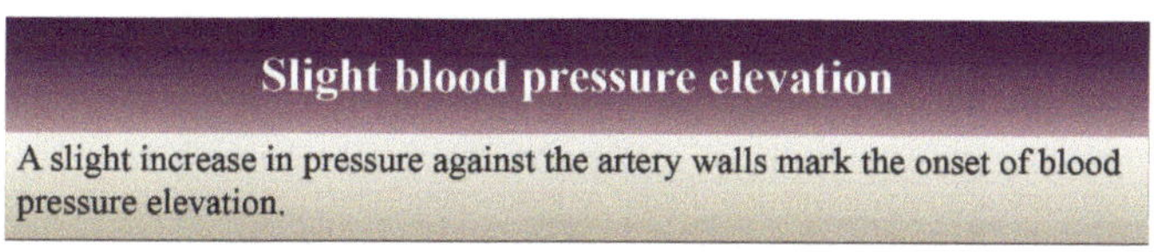

FIGURE 10.1

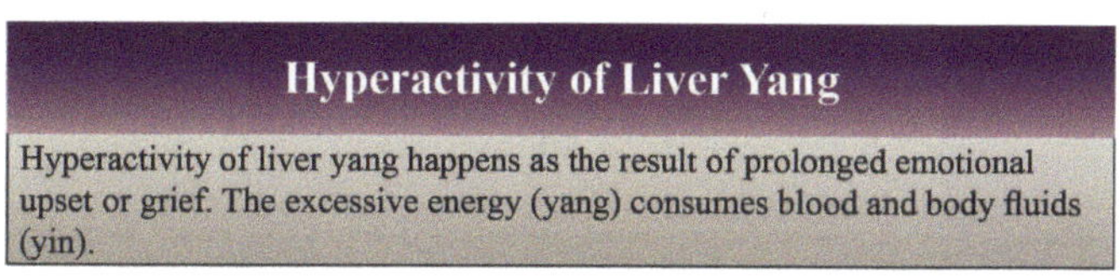

FIGURE 10.2

(a) Clinical presentation: headache, distended feeling in the head, vexation, moody and temperamental, flushed eyes, bitter taste in mouth, dizziness, tinnitus, restlessness.

Tongue = red with yellow coat

Pulse = wiry and forceful

See Figure 10.3.

(b) Treatment:

- Therapeutic principle: nourish yin, subdue yang
- Medicine: Gambirplant Branch Formula 3× daily between meals with warm water
- Exercise: Tai Qi and qigong 2× daily 15 minutes

(c) Medicated diet:

- Chrysanthemum flower tea 12 g, drink freely hot or cold as desired
- Mung bean and kelp soup 2× daily year round, avoid accompanying with greasy food
- Chrysanthemum and Hawthorn Rice Gruel 1× daily during breakfast in warm months

See Figure 10.4.

(a) Clinical presentation: headache, dizziness, tinnitus, restlessness, flushed face, 5 palm heat, menopause, vertigo, dry eyes, blurred vision, vexation, poor memory, aching pain in the loins and legs, and dry mouth.

Tongue = red with no fur

Pulse = fine at both kidney positions yet thready

See Figure 10.5.

(b) Treatment:

- Therapeutic principle: replenish and nourish yin, stabilize yang
- Medicine: Yang Hyperactivity Checking with 7 Drugs Formula 3× daily 5 day

Clinical Pearl:

Press on the lateral sides of the legs from hip to knee for positive tenderness/ticklish sensation, then also check/treat for a bacterial or viral infection. Formula: Modified Decoction of Gentiana for Purging Liver Fire 2x daily 5 days.

FIGURE 10.3

Deficiency of Liver and Kidney Yin

Liver stores blood (yin), kidneys store essence (yin) and they transform into each other systemically. Both are needed to help create and balance energy (yang) for the body. Deficiency of yin causes an excess of yang.

FIGURE 10.4

Clinical Pearl:

Have patient monitor urine color. There may be alternating dark yellow and normal/pale associated with racing thoughts, irritability, malaise and periods of relief. Needle LI11, then massage the head by pressing on yintang, Du20, Taiyang and GB20& 21 30 times for each. Circle over the entire scalp thirty times.

FIGURE 10.5

Essential Hypertension

The patient has an elevated supine, seated or standing blood pressure over two recent consecutive visits, and at least three essential hypertension risk factors.

FIGURE 10.6

Clinical Pearl:

Blood pressure fluctuates according to emotional factors, throughout the day, and at different times of the year. Sometimes it is difficult to get an accurate perspective on blood pressure even with very experienced health practitioners.

FIGURE 10.7

- Exercise: Tai Qi and Qigong 2× daily 15 minute
- Medicated diet:
  - Chrysanthemum flower tea: drink freely hot or cold as desired
  - Celery juice 250 g per day, drink freely
  - Onion and Chinese Date Soup 2× with non-greasy meals year round

STAGE 1 HYPERTENSION

Western Medicine

See Figure 10.6.

(a) Clinical presentation: the patient may have headaches, neck, back, shoulder or arm pain. In many cases, there are no other symptoms besides a high blood pressure, which is spotted during a physical examination.

See Figure 10.7

(b) Work up:
- Family history: overweight
- Hyperlipidemia: smoker
- Age: women >55, men >45: diabetes mellitus
- Chronic kidney disease: vision problems or retinopathy

(c) Treatment:
- Medicine: 1 or 2 medicine selections from Table 10c
- Increase physical activity: continuous movement 30 minutes per day
- Diet: use the 'Dietary Approaches to Stop Hypertension' (DASH) diet
- Weight maintenance: target as close as possible to ideal BMI
- Lifestyle: stop smoking: although smokers feel that it relaxes, nicotine increases heart rate. Reduce or avoid alcohol: drink in moderation or avoid drinking

Chinese Medicine Differentiation

See Figure 10.8.

(a) Clinical presentation: palpitation, insomnia, day dreaming, brain fog and forgetfulness.

Tongue = red, or white coat hiding color

Pulse = forceful in the right kidney position and relatively weak at the left

See Figure 10.9.

(b) Treatment:
- Therapeutic principle: nourish yin and subdue yang
- Medicine: liver subduing and Wind Stopping Formula 3× daily
- Exercise: stretching and Tai qi 2× daily for 20 minutes

Deficiency of Yin Leading to Hyperactivity of Yang

Early Stage

Now there is profound consumption of yin (blood, body fluids and essence). Kidney yin is unable to balance the energy of yang, causing a rapid and momentous rise.

FIGURE 10.8

Clinical Pearl:

Use diagnostic testing to monitor for signs of thyroid disorders and hyperaldosteronism.

FIGURE 10.9

(c) Medicated diet:
- Chrysanthemum flower tea 12 g, drink freely hot or cold as desired
- Chrysanthemum and Hawthorn Rice Gruel 2× daily during warm months
- Mung bean and kelp soup 2× daily year round; avoid accompanying with greasy food

See Figure 10.10.

(a) Clinical presentation: palpitation, insomnia, day dreaming, irritability and vexation, mental weariness, nocturnal emission, amnesia, dry stools.

Tongue = red or light purple with a little tongue coat

Pulse = thready or wiry and rapid in all positions

See Figure 10.11.

(b) Treatment:
- Therapeutic principle: nourish kidney yin and tonify heart
- Medicine: prescription #1 Yang Hyperactivity Check with 7 Drugs Formula #2 Blood Pressure Reducing Decoction and Modified Cardiotonic Pill, both 5× daily for 7 days
- Exercise: stretching and Tai qi 3× daily for 20 minutes

(c) Medicated diet:
- Hawthorn and Cassia Seed Tea 6× per day
- Chrysanthemum and Hawthorn Rice Gruel

See Figure 10.12.

Deficiency of Heart Yin

Mid Stage

Kidney yang has transformed into fire and flares upwards, and the heat begins consuming yin (fluids and blood), threatening the circulating volume throughout the body to the heart and lungs.

FIGURE 10.10

Clinical Pearl:

This patient may already be showing signs of heart failure and kidney/endocrine disorders, and may be indicated for pharmaceutical drug therapy with or without herbs. With drug/herbs used proceed with prescription #1. With herbs only use prescription #2.

FIGURE 10.11

Deficiency of Yin and Yang

Late Stage

The kidney fire eventually burns itself out resulting in a period of deficiency of yin and yang. This stage sets the continued cycle of damage between the kidneys and heart, possibly involving the lungs and other organs.

FIGURE 10.12

Clinical Pearl:

This patient may already be showing early signs of heart failure, and indicated for pharmaceutical drug therapy only. If herbs only are used proceed with prescription.

FIGURE 10.13

Secondary / Resistant Hypertension

The patient age is <30 or > 55, maintains a blood pressure over 160/100 mm Hg over two recent visits, and at least three secondary hypertension risk factors.

FIGURE 10.14

(a) Clinical presentation: palpitation, insomnia, irritability, apathy, renal failure, soreness of loins and knees, cold extremities, 5 palm heat.

Tongue = red or pale

Pulse = rapid yet fine

See Figure 10.13.

(b) Treatment:
- Therapeutic principle: nourish yin and restore yang
- Medicine: Gastrodia and Uncaria Formula 5× daily with warm water
- Exercise: stretching and Tai Qi 3× daily for 20 minutes

(c) Medicated diet:
- Chrysanthemum flower tea 12 g, drink freely hot or cold as desired
- Mung bean and kelp soup
- Chrysanthemum and Hawthorn Rice Gruel

STAGE 2 HYPERTENSION

Western Medicine

See Figure 10.14.

Clinical Pearl: Pseudohypertension

Some chronically ill patients with diabetes or kidney disease may actually have orthostatic or postural hypotension. Medication for hypertension in these patients may aggravate their condition. The Osler's maneuver can be used while taking a blood pressure. A positive sign reveals a palpable artery during cuff constriction.

FIGURE 10.15

Deficiency of Vital Energy and Yin

This stage features simultaneous or alternating symptoms related to yin and yang deficiency. These include function failures of the other organs and coldness of skin and limbs, due to yang failure to warm. Yin deficiency causes flashes of false heat known as 5 palm heat.

FIGURE 10.16

(a) Clinical presentation: despite medication, the blood pressure continues to rise to 160/100+. The patient may have sleep apnea, exhaustion, breathing difficulties and early heart failure symptoms. Findings may reveal medication or dietary non-compliance. In other cases illicit drug use. See Figure 10.15.

(b) Work up:
- Uncontrolled diet
- Overweight
- Age: $<$ age 30 or $>$ age 55
- Complications: diabetes, kidney disease, aldosteronism, renovascular disease, coarctation of aorta, thyroid disease or tumors

(c) Treatment:
- Medicine: two or three medicines from Table 7c
- Continue lifestyle modification: DASH diet, and moderate daily exercise

Chinese Medicine

See Figure 10.16.

(a) Clinical presentation: spontaneous sweating, hot palms and soles followed by cold limbs, dry mouth, oliguria with dark yellow urine, constipation, pale complexion, palpitations, insomnia, shortness of breath, tinnitus, dizziness, lassitude, coated tongue (may be red), headache, intolerance of the cold, poor appetite, loose stools, edema, puffy face, no desire to speak.

Clinical Pearl:

If this patient is referred in from a western medicine specialist, the patient may or may not be indicated for Chinese medicine. The patient must comply with the strategic process of weaning from western medicine to Chinese herbs.

FIGURE 10.17

Maladjustment of Chong and Ren Channels

Severely weak liver qi and kidney essence tributes to unique complications in females. Elderly women may experience hormonal disturbances. Young women may experience complicated menstruation problems or have peri-menopausal symptoms much earlier. In addition, some female patients can begin experiencing stage 1 hypertension very quickly.

FIGURE 10.18

Tongue = pale and tender
Pulse = weak and thready

See Figure 10.17.

(b) Treatment:
- Therapeutic principle: nourish yin and restore yang
- Medicine: Modified Gastrodia and Uncaria Formula 5× daily with warm water
- Exercise: stretching and Tai Qi 3× daily for 20 minutes

(c) Medicated diet:
- Tremella and Jew's Ear tea
- Sea Weed Soup
- Chrysanthemum and Hawthorn Gruel
- Water Chestnut and Jellyfish Soup

FEMALES WITH HYPERTENSION

Chinese Medicine

See Figure 10.18.

(a) Clinical presentation: palpitation, restlessness, insomnia, exhaustion, irritability, bloating between the sterno-costal region and knees, rapid weight gain, menstrual irregularity, amenorrhea, headaches, thyroid problems, general malaise, and fluctuations in blood pressure with high spikes followed by low drops.

Pulse = thready

See Figure 10.19.

Clinical Pearl:

Iodine test results usually show considerable deficiency or complete depletion. Supplementing with about 198-700mg per day with monitoring improves the effects of treatment.

FIGURE 10.19

TABLE 10.5

Medicine	Action
Thiazide diuretics	Reduces water
Beta Blockers	Heart beats slower and with less force
ACE inhibitors	Relaxes blood vessels
ARBs	Relaxes blood vessels
Calcium channel blockers	Relaxes blood vessels
Renin inhibitors	Slows the production of renin
Alpha blockers	Reduces nerve impulses that narrow blood vessels
Alpha-beta blockers	Reduces nerve impulses, slow the heartbeat
Central-acting agents	Prevents nervous system heart rate increase and blood vessel narrowing
Vasodilators	Prevents the arteries from constriction

(b) Treatment:
- Therapeutic principle: nourish yin and restore yang
- Medicine: Gu Chong Tang 3× daily with warm water
- Exercise: stretching and Tai Qi 3× daily for 20 minutes

(c) Medicated diet:
- Jia Wei Xiao Yao Tea 3× daily
- Mung bean and kelp soup
- Chrysanthemum and Hawthorn Rice Gruel

Medicines

A. Western medicine drugs
See Table 10.5.

B. Chinese medicine formulas

The formula ingredients, as well as possible modification with other formulas and herbs, can be found in the pathological listings (Chapter 19).

TABLE 10.6

Medicine	Action
Modified Gambirplant Branch Formula	Reduces hypertension, headache and dizziness
Modified Yang Hyperactivity Check with 7 Drugs Formula	Reduces hypertension, headache, tinnitus, and mildly diuretic
Modified Liver subduing and Wind Stopping Formula	Reduces hypertension, thyroid problems, hyper aldosterosim, & tranquilizes the mind
Modified Blood Pressure Reducing Decoction	Reduces blood pressure, dreaminess, insomnia, brain fog, chest oppression
Modified Qi Replenishing and Yin Nourishing Formula	Stablizes blood pressure while recovering from dryness, exhaustion and headache
Modified Cardiotonic Formula	Reduces blood pressure, nourishes blood
Diabetes, kidney failure, aldosteronism, renovascular disease, coarctation of aorta, thyroid disease or tumors	Reduces blood pressure in serious cases, and coordinates with pharmaceutical drugs

Modifications allow for practitioners to adjust a formula according to clinical presentation while maintaining a treatment course

See Table 10.6.

SUMMARY IN THE CONTEXT OF PATIENT PERSPECTIVES FOR PERSONALIZING MEDICINE

Health practitioners treating patients with hypertension may wish to consider information from the patient narrative, especially the times that the blood pressure rise is noticed during rest and activity. Integrative medicine is more indicated and with favorable results in the patient with pre-hypertension to stage 1. Often times the blood pressure fluctuations happen during night time if the patient is also experiencing palpitations. During the day, the rise may happen around meal times, and during walking and other physical and emotional efforts. Another key piece of information may include whether the patient also experiences insomnia with the blood pressure fluctuations while in bed at night, and somnolence, which interferes with activity during the day. This is a sign of the disharmony between the heart yang and kidney yin. Whether or not the patient is on Western medicine pharmaceutical drugs, a suggestion may be to introduce or modify the Chinese medicine formula(s) with appropriate yang-increasing herbs that also assist with stabilizing blood pressure during the day. This suggestion may also assist with unwanted blood pressure lows after metabolizing a drug. This stabilizing strategy can be used for several days to a couple of weeks at a time, as a test of appropriateness. For the right patient, it may give an immediate feeling of emotional and physical well being during that time. The patient may then be ready to organize diet and lifestyle modifications which can further personalize the treatment plan.

FURTHER READING

[1] Rueifang G, Jingfeng Z. Integrative medicine the essential hypertension stated [J]. Henan Tradit Chin Med 2013;33(2):248–9.

[2] Ziyin C. Clinical research of Chinese herbal compound recipe in the treatment of essential hypertension[J]. China J Chin Med 2013;28(1):107–8.

[3] Levy D, et al. Evidence for a gene influencing blood pressure on chromosome 17 genome scan linkage results for longitudinal blood pressure phenotypes in subjects from the Framingham heart study. Hypertension 2000;36:477–83.

[4] Lifton RP. Genetic dissection of human blood pressure variation: common pathways from rare phenotypes. Harvey Lect 2004–5;100:71–101.

[5] Simon DB, et al. Gitelman's variant of Bartter's syndrome, inherited hypokalaemic alkalosis, is caused by mutations in the thiazide-sensitive Na-Cl cotransporter. Nat Genet 1996;12:24–30.

[6] De Jong JC, et al. Functional expression of mutations in the human NaCl cotransporter: evidence for impaired routing mechanisms in Gitelman's syndrome. J Am Soc Nephrol 2002;13:1442–8.

[7] Cohen JC, et al. Multiple rare alleles contribute to low plasma levels of HDL cholesterol. Science 2004;305:869–72.

[8] Franklin SS, et al. Hemodynamic patterns of age-related changes in blood pressure. The Framingham Heart Study. Circulation 1997;96:308–15.

[9] Veith I. Huang Ti Nei Ching Su Wen. The Yellow Emperor's classic of internal medicine. Baltimore, Maryland, USA: Williams & Wilkins; 1949.

[10] Guyton AC. Abnormal renal function and autoregulation in essential hypertension. Hypertension 1991;18:49–53.

[11] Gitelman HJ, Graham JB, Welt LG. A new familial disorder characterized by hypokalemia and hypomagnesemia. Trans Assoc Am Physicians 1966;79:221–35.

[12] Province MA, et al. A meta-analysis of genome-wide linkage scans for hypertension: the National Heart, Lung and Blood Institute Family Blood Pressure Program. Am J Hypertens 2003;16:144–7.

[13] Weir MR, et al. Influence of race and dietary salt on the antihypertensive efficacy of an angiotensin-converting enzyme inhibitor or a calcium channel antagonist in salt-sensitive hypertensives. Hypertension 1998;31:1088–96.

[14] Jinglun H. Traditional Chinese medicine for cardiovascular diseases. Beijing, China: Academy Press; 1995.

PART III

Cardiac Rhythm Diseases

Chapter 11

Arrhythmia

Chapter Objectives

After studying this chapter you should be able to:

1. Explain the reasons why arrhythmia may or may not be indicated for Chinese medicine therapy
2. Describe some genetic determinants and mechanisms of arrhythmia
3. Discuss the general EKG lead findings for atrial fibrillation
4. Discuss the indications for Chinese medicine treatment in atrial fibrillation
5. Summarize the context of personalizing medicine

OVERVIEW

This chapter discusses aspects of arrhythmia for consideration of treatments using integrative care. Because several types of arrhythmia result in sudden death, a diagnosed patient often requires close monitoring using Western medicine methods.

Acute atrial fibrillation can present as a risk factor towards stroke. However, it may also result from a recent illness, or from emotional or temporary lifestyle factors. This type of arrhythmia may be suitable for treatment in the integrative medicine setting. With guidance in treatment, a patient may be able to take measures to prevent some of the more pathologic changes found in more serious atrial fibrillation. For example, chronic inflammation causes atrial fibrosis which leads to dilation. From there, conduction problems lead to a sequence of events towards heart failure.

Chinese medicine differentiates arrhythmia as blood stasis, qi and yin deficiency, blocking of the heart vessels, phlegm, yang deficiency, cold and dampness. It may be indicated as a prevention measure in patients presenting with acute atrial fibrillation. Close monitoring of arrhythmia – especially in more serious patients, those with other complicating diseases and rare familial genetic disorders – is especially crucial.

Integrative Cardiovascular Chinese Medicine. DOI: http://dx.doi.org/10.1016/B978-0-12-420014-2.00008-9

Key Points

- Atrial fibrillation may manifest as a result of inflammation, emotional or lifestyle factors at any age, but incidence commonly increases with age.
- Atrial fibrillation is a form of arrhythmia which may be indicated for treatment in Chinese medicine.
- Tachycardia is often differentiated as blood stasis and qi and yin deficiency.
- Bradycardia is usually differentiated as blood stasis which blocks the heart vessels, phlegm, yang deficiency, cold and dampness.
- Atrial fibrillation is not always caused by genetic mutations nor is it familial.
- Dilation of heart chambers not only causes blood pressure changes, it also activates RAAS, which leads to remodeling and more fibrosis in the myocardium and the SA and AV nodes. Diseases such as sick sinus syndrome can develop in predisposed patients.
- Cardiovascular diseases in which miRNA has contributed include valvular disease and atrial fibrillation.
- One mutation on a *KCNE2*, *KCNJ2* and *KCNQ1* gene involves the replacement of certain key amino acids in proteins which make up the channels that regulate potassium flow.
- The *KCNQ1*gene may be more involved with diseases such as ovarian cancer, type II diabetes and LQTS.
- Mechanisms of arrhythmia include disorders of impulse formation, disorders of impulse conduction and disorders of re-entry.
- II, III, aVF and V5 are leads to interpret.
- In atrial fibrillation, the P waves are usually hidden in the jagged waves.
- In atrioventricular re-entrant tachycardia or atrioventricular round-trip tachycardia the P wave is upside down, beginning immediately after the QRS wave group. There are extended waves during the PR intervals with the onset of tachycardia.
- Chest X-ray determines heart enlargement or pulmonary causes. Pneumonia may cause atrial fibrillation.
- Echocardiography determines the dimensions of the chambers, septum and valves in relationship to blood flow.
- Holter and event monitors record electrical signals during a period of 24 to 48 hours.
- Tilt table testing determines the cause of fainting spells.
- Stress test determines the magnitude of symptoms during stress.
- Electrophysiology study (EPS) determines the cause of more serious arrhythmias.

- An implantable loop recorder determines the cause of serial syncope and palpitations by monitoring abnormal heart rhythms over 12–24 months.
- Coronary angiography determines the magnitude of coronary artery block. A catheter is inserted through an artery in the arm leg or neck towards the coronary arteries. A contrast dye is injected, providing a view of the artery condition on X-ray.

PATHOGENESIS

Chinese Medicine

In Chinese medicine, the approaches to the treatment of types of arrhythmia is limited. Two main reasons involve the unknown factors in current research. One is that ventricular arrhythmia is often identified, post mortem, as the cause of sudden death in otherwise healthy individuals. The other is that the identification of types which lead to sudden death has not led to conclusive methods to prevent it. Arrhythmia is categorized in Chinese medicine as xiong bi (shee-ong bee)/chest obstruction, xin ji (sheen jee)/palpitation and xuan yun/vertigo. Atrial fibrillation is a form of arrhythmia which may be indicated for treatment in Chinese medicine. It has genetic causes in some patients, and may be a temporary symptom of emotional and lifestyle problems in others. Arrhythmia is considered a deficiency manifesting as tachycardia and bradycardia. Tachycardia is often differentiated as blood stasis and qi and yin deficiency. Bradycardia is usually differentiated as blood stasis which blocks the heart vessel, phlegm, yang deficiency, cold and dampness.

Western Medicine

Atrial fibrillation may manifest as a result of inflammation, emotional or lifestyle factors at any age, but incidence commonly increases with age. Other cardiovascular risk factors which increases with age include blood pressure, valvular diseases, lipoprotein disorders, renal disorders and diabetes mellitus. These factors have been shown to have a high incidence in the world's population as a genetic predisposition towards their incidence. It could be possible to consider atrial fibrillation as a risk factor based on genetics. However, atrial fibrillation is not always caused by genetic mutations, nor is it familial.

Some pathologic changes found in atrial fibrillation are due to inflammation, fibrosis and dilation. Inflammation causes fibrosis, and fibrosis causes dilation of the atria. Dilation of heart chambers not only causes blood pressure changes, it also activates RAAS – which leads to remodeling and more fibrosis in the myocardium and the SA and AV nodes.

The known factors involved in atrial fibrillation include:

- Heredity: arrhythmia can occur in families through genetics, emotional or lifestyle issues
- Environmental: stress, caffeine and alcohol consumption
- Demographic: age, race, gender
- Health state: hypertension, rheumatic heart diseases, inflammation and fibrosis within the atrial tissues, endocrine disorders, valvular diseases, cardiomyopathy, pulmonary diseases (sarcoidosis, pneumonia, embolism) neurologic diseases involving hemorrhage and ischemia, and post-operative use of certain medications

DETERMINANTS AND MECHANISMS

Genetic Markers

1. miRNA. MicroRNA are involved in many biologic responses and processes. It regulates the expression of genes throughout the body, including the cardiovascular system. In the cardiovascular system, miRNA expresses itself in the regulation of KEGG signaling channels of ions within and outside of cells. Some noted channels include calcium signaling pathways and Wnt signaling pathways. Cardiovascular diseases in which miRNA has contributed include valvular disease and atrial fibrillation.
2. KCN (potassium channel) gene mutations and familial atrial fibrillation (FAF). Three mutations on genes *KCNE2*, *KCNJ2* and *KCNQ1* involve the replacement of certain key amino acids in proteins which make up the channels that regulate potassium flow.

   KCNE2 gene mutation is the result of the amino acid arginine being replaced by cysteine. The channel remains open, allowing for unregulated flow of potassium during resting and action potentials. Valine, the amino acid which helps to make the *KCNJ2* gene is replaced by isoleucine, causing the mutation. The channel remains open, allowing for unregulated flow of potassium during resting and action potentials. The *KCNQ1* gene is not considered to be a common cause of FAF. The following genetic disorders are connected with FAF:
   - Romano–Ward syndrome/long QT syndrome. Potassium channels open slower and close faster than usual. The decreased outflow of potassium ions causes syncope and possible sudden death.
   - Short QT syndrome. The increased flow of potassium out of cells causes the irregular shortened interval between heartbeats, which may cause syncope and sudden death.
   - Andersen–Tawil syndrome. This condition causes arrhythmia marked by muscle weakness and paralysis due to mutations which change the shape of the channel, causing potassium transport problems.

- Jervell and Lange–Nielsen syndrome. The KCNQ1 has fewer, shorter proteins and is unable to help build potassium channels, which causes congenital arrhythmia and deafness.
- Long QT syndrome. Potassium channels are affected, causing polarization failure between the action and resting potential phases.

a. *KCNE2* gene. The proteins of this gene form the structure of the potassium voltage-gated channel, which transports potassium in and out of cardiac myocytes and maintains the normal rhythm. Cardiovascular disease involves a mutation in this gene:
  - Romano–Ward syndrome/long QT syndrome

b. *KCNJ2* gene. On this gene, the potassium inwardly-rectifying channels, are also responsible for generation of signals in the conduction system. These channels are found in cardiac and muscle tissue, and they assist with contraction and relaxation. Cardiovascular diseases which feature a mutation in this gene:
  - Short QT syndrome
  - Andersen–Tawil syndrome

c. *KCNQ1* gene. This gene and the KCNE1 are part of the components of potassium channels. Channels with this protein regulate cardiac rhythm, are found in the ear for balance and hearing, and are involved in lung, kidney and gastrointestinal functioning. In addition, this gene has been identified in sudden death syndrome (SIDS) and is also triggered during menopause with diseases such as ovarian cancer, type II diabetes and LQTS. Cardiovascular diseases involve a mutation in this gene:
  - Jervell and Lange–Nielsen syndrome
  - Romano–Ward syndrome/long QT syndrome (LQTS)
  - Short QT syndrome

3. ACE gene polymorphism. Patients with this gene may experienced altered responses to ACE inhibitor medication.

Constitutional Factors

See Table 11.1.

Mechanisms

Disorders of Impulse Formation

Ectopic rhythms causing abnormal automaticity which are noticed in the resting and action potentials: delayed after-depolarizations, early after-depolarizations and parasystole. An example is recognized in long QT syndrome.

Disorders of Impulse Conduction

Physiologic changes contributing to tachycardia and bradycardia: conduction blocks of rapid impulses (tachy), blocks in the AV node (decremental-dependent block) re-pacing tachycardia (entrainment).

TABLE 11.1

Constitution	General Incidence of Arrythmia
Infants and children	SIDS in infants, heart murmur, hearing and developmental problems
Teenage and young adult	Possible age of onset and quality of menstruation in females, for future reference to gestational arrythmia and genetic passing of conduction problems. Consider cardiopulmonary factors involved in endurance problems during rigourous activities such as sports
Middle-age	Female menopause related conduction problems, lipid disorders and cancers. Andropause related problems in men
General geriatric	Arrythmia incidence elevating to 80% with various symptoms of tachycardia and possible medication induced bradycardia

Disorders of Re-Entry

SA nodal cells with delayed firing, and contraction of the atria between the phases of the action potential causes abnormal rhythms, such as atrial flutter and atrial fibrillation. Further delays in the AV nodal firing lead to abnormal rhythms such as ventricular tachycardia, ventricular cardiomyopathy and ventricular fibrillations.

EXAMINATION OF THE PATIENT

Physical Examination

- Check for leg, ankle and foot edema, which can indicate heart failure
- Inquire and palpate for thyroid changes

a. Blood pressure:
   - This is taken in the sitting, standing or supine position, and readings indicate stages of hypertension if present during the assessment.

b. Pulse in TCM:
   Chinese pulse diagnosis:
   - Rapid = true heat, yin deficiency heat, infection
   - Weak = deficiency
   - Wiry = hyperactive liver condition.

Laboratory Tests

Routine Tests

1. 12-lead electrocardiogram. This test is used to identify certain patterns which provide evidence of arrhythmias, such as P wave duration and

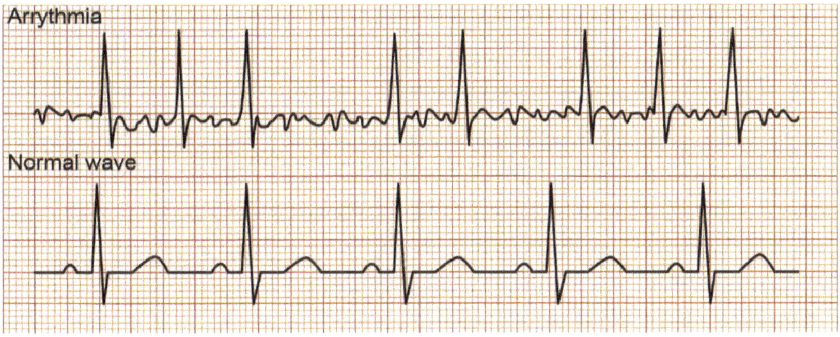

FIGURE 11.1

f waves, rhythm, branch blocks, pre-excitation, left ventricular hypertrophy, prior myocardial infarctions and various other forms of arrhythmia.

a. Atrial fibrillation: the following information represents a brief overview of the condition of the PQRS waves noticed in arrhythmia. Variations during presentations should be considered.
   See Figure 11.1.
   - The P wave: P waves are usually hidden in the f waves. Impulse rate is 350–600/minute. The QRS wave may be wide or may maintain normal sinus rhythm.
b. Paroxysmal supraventricular tachycardia:
   - The patient may complain that heart sensation is thumping against the chest wall. Impulse rate is 150–250/minute. The QRS wave may maintain normal sinus rhythm.
c. ST segment depression or t wave inversion: leads: II, III, aVF and V5
   Atrioventricular re-entrant tachycardia:
   - The P wave is upside down, beginning immediately after the QRS wave group
   - Extended waves during the PR intervals with the onset of tachycardia

   Accessory pathways formed in ventricular tachycardia:
   - The P wave is upside down and is located in between each QRS wave group, tips left concealed.

2. Transthoracic echocardiogram. This test identifies the condition, size and dimensions of the left and right atria, the left and right ventricles, valves, and the presence of thrombus.

Other Tests

a. Chest X-ray: determines heart enlargement or pulmonary causes. Pneumonia may cause atrial fibrillation.

b. Holter and event monitors: record electrical signals during a period of 24–48 hours.
c. Tilt table testing: determines the cause of fainting spells by lying on a table which moves from a downward to an upright position. The change in position may cause the patient to faint. The blood pressure, EKG and heart rate are monitored.
d. Stress test: determines the magnitude of symptoms during stress. The patient is monitored using PE, nuclear heart scanning or echocardiography during exercise or given medicine to cause faster heart rhythms. Symptoms to consider:
   - a very rapid heart rate due to pulmonary causes or ischemia
   - normal activity, but the patient develops shortness of breath with exertion
e. Electrophysiology study (EPS): determines the cause of more serious arrhythmias. An electrode wire is passed through an artery in the arm or leg to the heart to monitor cardiac conduction.
f. Implantable loop recorder: determines the cause of serial syncope and palpitations by monitoring abnormal heart rhythms over 12–24 months.
g. Coronary angiography: determines the magnitude of coronary artery block. A catheter is inserted through an artery in the arm, leg or neck towards the coronary arteries. A contrast dye is injected, providing a view or the artery condition on X-ray.

Blood Tests

See Table 11.2.

ACUTE ATRIAL FIBRILLATION

Western Medicine

a. Clinical presentation:
   - chest: discomfort, tightness, pain, palpitations or fluttering
   - lungs: breathing difficulties during normal activities or rest
   - fatigue and lack of energy
   - dizziness
b. Work up:
   - Family or non-history of atrial fibrillation
   - Age: any age
   - Race: any race, although possible differences in incidence were seen recently between women of European, Asian and African ancestry
   - Gender: both sexes
   - Lifestyle: smoker, caffeine and alcohol consumption
   - Thyroid problems
   - Infection: viral or bacterial

TABLE 11.2

Lab	Detections
Thyroid panel	Tests TSH/ T4 levels
Liver panel:	
Alanine aminotransferase (ALT)	Detects hepatitis
Alkaline phosphatase (ALP)	Increases during blockage of bile ducts
Aspartate aminotransferase (AST)	Detects damage to heart and muscle tissue
Bilirubin	Detects liver damage
Total Protein (TP)	Measures the liver derived proteins which affect the kidneys
	Measures blood proteins
Gamma glutamyl transferase (GGT)	Measures liver damage
Lactate dehydrogenase (LDH)	Measures cell damage throughout the body
Prothrombin time (PT)	Measures blood clotting and coagulation in relation to liver damage
Kidney function tests:	
Urinary albumin	Urinary albumin: Detects and monitors levels in patients with nephropathies
	Plasma renin: detects and monitors levels in patients with hypokalemia or chronic kidney disease
	Blood urea nitrogen (BUN): Detects and measures in patients with kidney and heart failure
Plasma renin	Hypokalemia and acute or chronic kidney disease
Blood urea nitrogen (BUN)	Measures the amount of urea nitrogen that is in the blood

- Cardiovascular diseases: hypertension, cardiomyopathy, coronary artery disease (CAD), valvular diseases, pulmonary diseases, congenital diseases

This type of patient, with mild symptoms, is usually evaluated with electrocardiogram (EKG) and echocardiogram (EEG). *Pharmaceutical medication may or may not be given.*

c. Treatment. ACC/AHA/ESC 2006 guidelines for the management of patients with atrial fibrillation. The recommendations classify the conditions for providing treatment from beneficial to harmful. The weight of evidence scrutinizes treatments based on the reliability of the source of information.

See Tables 11.3 and 11.4.

TABLE 11.3

Classification	Recommendations
Class I:	Conditions for which there is evidence and/or general agreement that a given procedure/therapy is beneficial, useful, and effective
Class II:	Conditions for which there is conflicting evidence and/or a divergence of opinion about the usefulness/efficacy of performing the procedure/therapy
Class IIa:	Weight of evidence/opinion is in favor of usefulness/efficacy
Class IIb:	Usefulness/efficacy is less well established by evidence/opinion
Class III:	Conditions for which there is evidence and/or general agreement that a procedure/therapy is not useful or effective and in some cases may be harmful

TABLE 11.4

Level of Evidence	Weight of Evidence
Level of Evidence A:	Data derived from multiple randomized clinical trials or meta-analyses
Level of Evidence B:	Data derived from a single randomized trial, or nonrandomized studies
Level of Evidence C:	Only consensus opinion of experts, case studies, or standard-of-care

Chinese Medicine

Tachycardia is often differentiated as blood stasis and qi and yin deficiency. See Figure 11.2.

a. Clinical presentation: emotional withdrawal, vexation, mania, palpitations, insomnia, dizziness, shortness of breath, and spontaneous sweating
Tongue = no significance
Pulse = wiry

b. Treatment:
- Therapeutic principle: nourish yin, subdue yang, clear heat
- Medicine: 3 × daily between meals with warm water
  - An shen ning xin formula

Hyperactivity of Liver Yang

The liver is associated with psycho-emotional factors involved with anger. Deficient liver yin causes the yang to easily rise in emotional situations.

FIGURE 11.2

Clinical Pearl:

This differentiation is commonly seen in acute atrial fibrillation caused by stress and other psycho-emotional factors.
Generally more common in normal sinus rhythm.

FIGURE 11.3

Blood Deficiency

Often after a long illness such as in a viral or bacterial infection, the deficient blood can produce symptoms of arrhythmia.

FIGURE 11.4

Clinical Pearl:

This differentiation is also commonly seen in myocarditis, cardiomyopathy, rheumatic heart disease, tachycardia, and atrial fibrillation.

FIGURE 11.5

- Si Ni decoction
- Wendan decoction
- Linggui Shugan decoction
- Xuefu Zhuyu decoction
- Taohong Siwu decoction

See Figures 11.3 and 11.4.

a. Clinical presentation: tiredness, listlessness, emotional withdrawal, palpitations, recent recovery from infection or illness, insomnia, coldness – especially limbs.
   Tongue = pale
   Pulse = rapid and weak
   See Figure 11.5.

TABLE 11.5

Medicine to control heart rate	Acute heart rate control setting
Esmolol: Class I, evidence C	For patients without accessory pathway
Metroprolol: Class I, evidence C	
Propanolol: Class I, evidence C	
Diliatazem: Class I, evidence B	
Verapamil: Class I, evidence B	
Amiodarone: Class IIa, evidence C	For patients with accessory pathways
Digoxin: Class I, evidence B	For patients with heart failure without accessory pathway
Amiodarone: Class I, evidence C	

b. Treatment:
- Therapeutic principle: nourishing blood, soothe the spirit
- Medicine: 3 × daily with warm water
  - Zhi gancao decoction
  - guipi decoction
  - fu mai decoction
  - Shen Fu decoction

MEDICINES

A. Western Medicine Drugs

See Table 11.5.

B. Chinese Medicine Formulas

See Table 11.6.

SUMMARY IN THE CONTEXT OF PATIENT PERSPECTIVES FOR PERSONALIZING MEDICINE

The treatment of arrhythmia has been a controversial subject for many years. The 2006 and 2011 ACC/AHA/ESC guidelines have helped health practitioners evaluate and manage treatments. Currently, there is no section which considers Chinese medicine as a possible non-pharmaceutical treatment for any type of arrhythmia.

TABLE 11.6

Medicine	Action
An shen ning xin Formula	Soothes the spirit
Fu mai decoction	Qi and blood deficiency
Shen Fu decoction	Recovers after illness, warms yang
Si Ni decoction	Warms coldness
Wendan decoction	Regulates qi and transforms phlegm
Linggui Shugan decoction	Phlegm retention
Xuefu Zhuyu decoction	Blood stagnation and stasis
Taohong Siwu decoction	Invigorate blood
Zhi gancao decoction	Helps recover from illness, regulates heart functioning
Guipi decoction	Nourishes blood

Current studies present novel methods for determining the cause and course of arrhythmia at the genetic level. Some studies conclude that general results are fit for guiding the treatments for all, while others point out specific cultural and racial differences on the genetic level which complicate generalizations. Integrative cardiovascular Chinese medicine focuses more on prevention, which mostly means attempting to provide early solutions which may help to delay or divert away from more aggressive forms of therapy if possible.

FURTHER READING

[1] Zhao Z, Shi W, Wang X. Classification of antiarrhythmic drugs, progress and security applications. Clin Med J 2007;5(3): Beijing.

[2] Qian M. Research progress of antiarrhythmic drugs. Contemp Med 2010;16(4):2010 Issue No.195, Guangdong.

[3] Wang J. Several methods of traditional Chinese medicine in treating arrhythmia. Peoples Mil Doct 1985;(3): Nanjing.

[4] Han Z, Qi J. Progress in study on mechanism of antiarrhythmic effect of traditional Chinese medicine. ITCM 2007 2007;24(2): Heilongjiang.

[5] Fang W. Antiarrhythmic clinical application of Chinese medicine. Chin Clin Doct 2001;29(4).

[6] Yang B-f, Cai B-z. Advances in the study of arrhythmogenic mechanisms. J Int Pharm Res 2010;37(2): Heilongjiang.

[7] Zhu H. Research progress on TCM treatment of arrhythmia. JETCM 2008;17(6): Wuhan.

[8] Xue X, Guo D, Sun H, et al. Wenxin Keli suppresses ventricular triggered arrhythmias via selective inhibition of late sodium current. Pacing Clin Electrophysiol 2013;36(6):732–40 doi: 10.1111/pace.12109.
[9] Fuster V, Rydén LE, Cannom DS, et al. ACC/AHA/ESC (2006) Guidelines for the Management of Patients with Atrial Fibrillation: a report of the American College of Cardiology/American Heart Association Task Force on Practice Guidelines and the European Society of Cardiology Committee for Practice Guidelines (Writing Committee to Revise the 2001 Guidelines for the Management of Patients With Atrial Fibrillation): developed in collaboration with the European Heart Rhythm Association and the Heart Rhythm Society. Circulation 2006;114(7):e257–354.
[10] Levy S. Classification system of atrial fibrillation. Curr Opin Cardiol 2000;15(1):54–7.
[11] Leung RS, Huber MA, Rogge T, et al. Association between atrial fibrillation and central sleep apnea. Sleep 2005;28(12):1543–6.
[12] Fitzmaurice DA, Hobbs FD, Jowett S, et al. Screening versus routine practice in detection of atrial fibrillation in patients aged 65 or over: cluster randomised controlled trial. BMJ 2007;335(7616):383.
[13] Mant J, Fitzmaurice DA, Hobbs FD, et al. Accuracy of diagnosing atrial fibrillation on electrocardiogram by primary care practitioners and interpretative diagnostic software: analysis of data from screening for atrial fibrillation in the elderly (SAFE) trial. BMJ 2007;335(7616):380.
[14] Zipes Ziad F, Issa John M, Miller Douglas P. Clinical arrhythmology and electrophysiology: a companion to Braunwald's heart disease. Philadelphia: Saunders; 2009.
[15] Blackshear JL, Odell JA. Appendage obliteration to reduce stroke in cardiac surgical patients with atrial fibrillation. Ann Thorac Surg 1996;61(2):755–9.
[16] Conen D, Tedrow UB, Cook NR, et al. Alcohol consumption and risk of incident atrial fibrillation in women. JAMA 2008;300(21):2489–96.
[17] Sweeney MO, Bank AJ, Nsah E , et al. Minimizing ventricular pacing to reduce atrial fibrillation in sinus-node disease. N Engl J Med 2007;357(10):1000–8.
[18] Fox CS, Parise H, D'Agostino RB, et al. Parental atrial fibrillation as a risk factor for atrial fibrillation in offspring. JAMA 2004;291(23):2851–5.
[19] Saffitz JE. Connexins, conduction, and atrial fibrillatio. N Engl J Med 2006;354(25): 2712–4.
[20] Shimizu W. Atrial fibrillation and genetic abnormalities. Nihon Rinsho 2013;71(1): 161–1666.
[21] Klabunde R. Cardiovascular physiology concepts. Baltimore: Lippincott Williams & Wilkins; 2005ISBN 978-0-7817-5030-1.

Chapter 12

Sick Sinus Syndrome

Chapter Objectives

After studying this chapter you should be able to:

1. Explain the integrative medicine perspective on the connection between atrial fibrillation and sick sinus syndrome
2. Describe some genetic determinants and mechanisms of sick sinus syndrome
3. Discuss the characteristic EKG finding for sick sinus syndrome
4. Discuss the pattern for preventive treatment using integrative Chinese medicine
5. Summarize the context of personalizing medicine

OVERVIEW

This chapter outlines the treatment of sick sinus syndrome as an early prevention strategy.

In Chapter 11, the perspective of atrial fibrillation was discussed from a prevention perspective. As more of a temporary symptom in some patients, atrial fibrillation can be due to a recent infection, inflammation, illness, or psycho-emotional and lifestyle factors. The long-term effects from a chronic condition include fibrosis – which can lead to atrial dilation. From there, conduction problems give rise to a sequence of events leading to sick sinus syndrome in predisposed individuals later in life.

Chinese medicine differentiates both atrial fibrillation and sick sinus syndrome similarly. However, the patterns which lead to the end stages in sick sinus syndrome are as follows: heart yang deficiency, deficiency of heart and spleen, qi deficiency and blood stasis, and deficiency of yin and yang. Long-term patterns, which develop into a chronic condition, either result in the condition or activate a genetic predisposition. It is notable that many prescription medications which treat the conduction system and other cardiovascular disorders also cause complications such as sick sinus syndrome.

Integrative Cardiovascular Chinese Medicine. DOI: http://dx.doi.org/10.1016/B978-0-12-420014-2.00009-0

Key Points

- Atrial fibrillation may manifest at any age as a result of inflammation, or emotional or lifestyle factors, but incidence commonly increases with age. Sick sinus syndrome may be an end-stage condition in arrhythmia.
- Elderly patients with sick sinus syndrome may exhibit signs of fibrosis, inflammation and alternating tachyrhythms and bradyrhythms.
- Sick sinus syndrome is differentiated into qi deficiency and blood stasis, deficiency of heart and spleen, heart yang deficiency, fire excess from yin deficiency and deficiency of yin and yang.
- Atrial fibrosis is a complication in atrial fibrillation which may cause sick sinus syndrome.
- HCN channels are involved in the automaticity of the SA node, and mutations may be involved with sinus bradycardia.
- Middle-aged patients should be concerned about arrhythmia, especially if the patient has a history of hypertension around 140/90. Prevention methods using integrative Chinese medicine may help delay or avoid a chronic condition.
- The differentiations in this section are for following and assessing disease progression and determining the differences between the indication of Chinese medicine and referral of the patient for a pacemaker.
- The prevention window may likely be more of a delay than avoidance of the condition in some patients.
- During the prevention period, the treatment principle is to balance the heart and kidneys and tonify the spleen. This helps to nourish yin, move blood, tonifying yang and qi.
- End stages may necessitate more palliative care.

PATHOGENESIS

Chinese Medicine

In Chinese medicine, sick sinus syndrome is classified under palpitations, and is considered a health state with symptoms of nearly depleted innate essence. As the conduction system declines, the heart becomes weaker, over all energy diminishes. The mood and personality can decline into anxiety. Sick sinus syndrome is differentiated into qi deficiency and blood stasis, deficiency of heart and spleen, heart yang deficiency, fire excess from yin deficiency and deficiency of yin and yang.

Western Medicine

Sick sinus syndrome is a disorder of the conduction system and characterized by alternating periods of tachycardia and bradycardia. Though it is considered rare, incidence is usually found among the elderly. It is associated with several types of bradycardia, tachycardias and atrial fibrillation. Atrial fibrosis is a complication in atrial fibrillation which may cause sick sinus syndrome.

The known factors involved in sick sinus syndrome include:

- Heredity: sick sinus syndrome was once thought to be an acquired condition, but has been shown to occur in families
- Demographic: age, gender
- Health state: hypertension, rheumatic heart diseases, inflammation and fibrosis within the atrial tissues, endocrine disorders, valvular diseases, cardiomyopathy, pulmonary diseases (sarcoidosis, pneumonia, embolism) neurologic diseases involving hemorrhage and ischemia, and post-operative use of certain medications.

Determinants and Mechanisms

Genetic Markers

HCN4 mutations. HCN channels are involved in the automaticity of the SA node, and mutations may be involved in the abrupt switch to sinus bradycardia from tachycardia, which is responsible for reduced excitability and is characteristic in sick sinus syndrome.

Constitutional Factors

See Table 12.1.

- Mechanisms: similar to arrhythmia/atrial fibrillation

TABLE 12.1

Constitution	General Incidence of Sick Sinus Syndrome
Teenage and young adult	Extremely rare incidence in childhood and young adult. Congenital atrial fibrillation may be a concern in those patients. Family history and genetic profiling may help with avoiding certain risk factors which may activate symptoms later in life. Pre-hypertensive patients should caution themselves with prevention measures
Middle-age	Symptoms of arrhythmia especially if the patient has a history of hypertension around 140/90. Prevention methods using integrative Chinese medicine may help delay or avoid a chronic condition
General geriatric	Elderly patients may be exhibiting signs of fibrosis, inflammation and alternating tachy and bradyrhythms

- Disorders of impulse formation. Ectopic rhythms causing abnormal automaticity which are noticed in the resting and action potentials: delayed after-depolarizations, early after-depolarizations and parasystole. An example is recognized in long QT syndrome
- Disorders of impulse conduction. Physiologic changes contributing to tachycardia and bradycardia: conduction blocks of rapid impulses (tachy), blocks in the AV node (decremental-dependent block) re-pacing tachycardia (entrainment)
- Disorders of re-entry. SA nodal cells with delayed firing, and contraction of the atria between the phases of the action potential causes abnormal rhythms, such as atrial flutter and atrial fibrillation. Further delays in the AV nodal firing leads to abnormal rhythms such as ventricular tachycardia, ventricular cardiomyopathy and ventricular fibrillations.

EXAMINATION OF THE PATIENT

Visual Observation

Perform a neurologic examination for possible stroke symptoms by noticing the patient's gait upon entering the room and their bilateral grip strength.
Listening:

- Head: distended feeling, insomnia, day dreaming, brain fog and forgetfulness and dizziness
- Mouth: pale tongue color
- Emotions: vexation, moody and temperamental or sullen and listless
- Limbs: pain, coldness
- Others: hormonal problems, climacteric events, sexual problems

Physical examination:

a. Blood pressure:
    - This is taken in a sitting, standing or supine position, and readings indicate hypotension and hypertension
b. Pulse in TCM:
    - Chinese pulse diagnosis:
        - Weak = deficiency
        - Thready = hyperactive kidney (yin or yang)

Laboratory Tests

Routine Tests: EKG

See Figure 12.1.
Atrial fibrillation (Chapter 11).
See Figure 12.2.

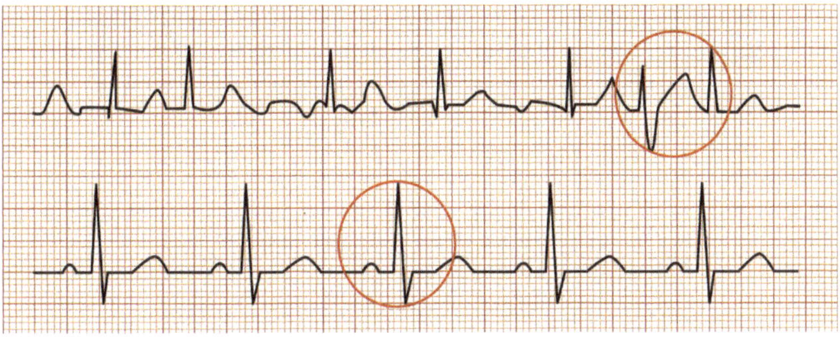

FIGURE 12.1

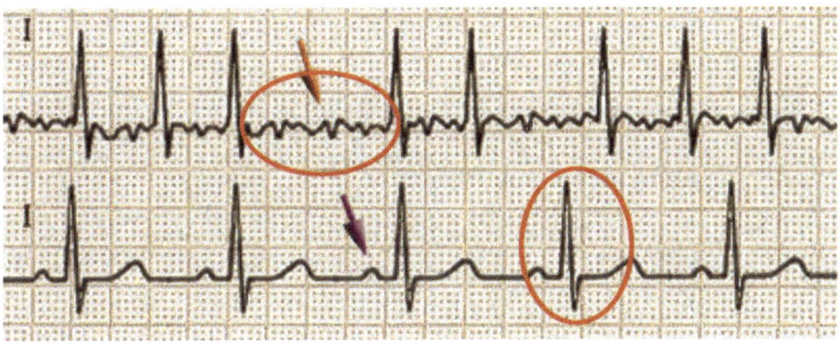

FIGURE 12.2

CLINICAL TREATMENT

Western Medicine

a. Clinical presentation. Some patients lack symptoms while others may present with tiredness, headache, nausea, chest pain, palpitations and syncope.
b. Work up and treatment. Close monitoring of this patient requires regular EKG assessment. Results may reveal sinus bradyarrhythmia, sinus node re-entrant tachycardia, and a prolonged asystolic period after tachycardia.

A pacemaker is usually implanted to control the tachycardia by maintaining the heart rate around 50 bpm. Medications may or may not be given for this patient.

Chinese Medicine

The differentiations in this section are for following and assessing disease progression and determining the differences between the indication of Chinese medicine and referral of the patient for the pacemaker.

See Figure 12.3.

Heart Yang Deficiency

Heart yang deficiency includes kidney yang deficiency. The main signs in this syndrome include insomnia, hot upper half of the body and cold lower half.

FIGURE 12.3

Clinical Pearl:

This may be the crucial period for a patient to balance the heart and kidneys and work on nourishing yin, tonifying yang and qi.

FIGURE 12.4

Deficiency of Heart and Spleen

The relationship between the heart and spleen involves generation and circulation of blood. The signs of blood deficiency manifest in the dizziness, weakness and lusterless complexion.

FIGURE 12.5

a. Clinical presentation: empty heart sensation, choking chest sensation which worsens with movement, pale complexion, cold limbs, sore and weak lower back, lower limb edema.
- Tongue = pale
- Pulse = weak and thready

See Figure 12.4.

b. Treatment:
- Therapeutic principle: warm heart yang, relax the shen
- Medicine: shen fu decoction
  - gui zhi tang
  - you gui zhen zhu decoction

A tea made using ginseng and wild jujube can be drunk freely as a palliative beverage if the patient will tolerate it. This decoction warms the body and invigorates yang.

See Figure 12.5.

a. Clinical presentation: heart palpitations, weakness, dizziness, syncope, shortness of breath, amnesia, lusterless complexion.
- Tongue = pale red
- Pulse = weak

See Figure 12.6.

Clinical Pearl:

This may be a final period in treatment where Chinese herbal medicine may have blended with Western medicine with poor results. The patient may be indicated for pharmaceutical drugs only.

FIGURE 12.6

Qi Deficiency and Blood Stasis

Patient begins to feel weaker with chest pain and palpitations coming more frequently.

FIGURE 12.7

Clinical Pearl:

This patient may establish the most serious symptoms leading to weaning off pharmaceutical medications and preparing for a pacemaker.

FIGURE 12.8

b. Treatment:
- Therapeutic principle: replenish qi, invigorate the spleen, relax the shen
- Medicine: gui pi tang
  - gan cao tang

A tea made using ginseng and wild jujube can be drunk freely as a palliative beverage if the patient will tolerate it. This decoction warms the body and invigorates yang.

See Figure 12.7.

a. Clinical presentation: heart palpitations with choking sensation in the chest, chest pain which worsens at night.
- Tongue = dark purple with ecchymosis
- Pulse = weak

See Figure 12.8.

b. Treatment:
- Therapeutic principle: invigorate qi, nourish blood, warm heart yang
- Treatment: tui na

c. Medicated diet:
- A tea made using ginseng and wild jujube can be drunk freely as a palliative beverage if the patient will tolerate it. This decoction warms the body, nourishes blood and invigorates yang

See Figure 12.9.

Deficiency of Yin and Yang

Yin and yang are unable to perform their duties of mutual promotion and restriction, as the deficiency of both leads eventually to end stage separation.

FIGURE 12.9

Clinical Pearl:

Elderly patients in this category may have a steady history of hospitalization and may be indicated for palliative care.

FIGURE 12.10

TABLE 12.2

Medicine	Action
Gui zhi tang	Warms to body
Shen fu decoction	Recovers after illness, warms yang
Gui pi tang	Nourishes and tonifies blood
Zhi gan cao tang	Helps recover from illness, regulates heart functioning
You gui zhen zhu decoction	Tonifies kidney yin and calms the spirit

a. Clinical presentation: palpitations, insomnia, chest tightness, shortness of breath, dizziness, ear ringing, cold limbs and body chills.
   - Tongue = pale
   - Pulse = fine

   See Figure 12.10.
b. Treatment:
   - Therapeutic principle: nourish yin, tonify yang
   - Treatment: tui na
c. Medicated diet: a tea made using ginseng and wild jujube can be drunk freely as a palliative beverage if the patient will tolerate it. This decoction invigorates yang. Modify to add other yin nourishing herbs.

MEDICINES

Chinese Medicine Formulas

See Table 12.2.

SUMMARY IN THE CONTEXT OF PATIENT PERSPECTIVES FOR PERSONALIZING MEDICINE

Health practitioners treating patients with hypertension may wish to consider the recorded patient history, especially the visits and stays as an in-patient. This information can explain the patterns and progression and give insight about the seriousness of the patient's condition as a reference.

Like the treatment suggestion in Chapter 13 (angina pectoris), patients who benefit the most from using Chinese medicine treatments alone are those who may have not experienced a life-threatening episode which has debilitated the patient towards the end stage. Integrating Chinese and Western medicine may be indicated on the hospital inpatient level where close monitoring can guide more favorable results during the course of treatment. On a base level in treatment, ren shen (ginseng), alone or featured in formulas, has the marvelous ability to rescue a patient on the brink of poor prognosis, and may improve and tonify the deficiency of yin and yang.

FURTHER READING

[1] Yan G, Renguang L, Fengqin Z. Research status of sick sinus syndrome. J Liaoning Med Univ 2012;33(2):154–7.

[2] Xin-sheng W. TCM treating in differentiation of sick sinus syndrome. Wang Xin-sheng Clin J Chin Med 2011;3(4):74.

[3] Ping WK. J Electrocardiol 2003;22(4).

[4] Yan G, Renguang L, Fengqin Z. Research status of sick sinus syndrome. J Liaoning Med Univ 2012;33:2.

[5] Xinsheng W. 4 TCM treating in differentiation of sick sinus syndrome. Clin J Chin Med 2011;3(4).

[6] Yun-geng G. Atrial fibrillation in sick sinus syndrome cause treatment failure: a clinical observation. Fujian Med J 2011;33(3):23–5.

[7] Peisheng Z, Jianfeng Y, Xue L. Influence of permanent cardiac pacing therapy on the prognosis in the patients with sick sinus syndrome. J Pract Diagn Treat 2006;10:734–5.

[8] Benson DW, Wang DW, Dyment M, et al. Congenital sick sinus syndrome caused by recessive mutations in the cardiac sodium channel gene (SCN5A). J Clin Investig 2003;112(7):1019–28.

PART IV

Chamber and Valve Diseases

Chapter 13

Angina Pectoris

Chapter Objectives

After studying this chapter you should be able to:

1. Explain the treatment limitations of using integrative Chinese medicine
2. Describe the mechanisms of angina pectoris
3. Describe the role of Chinese medicine during the indication stage
4. Discuss treatment methods in Chinese medicine
5. Summarize the context of the patient's role in personalizing medicine

OVERVIEW

This chapter describes the stages of angina pectoris and the possibility of integrated medicine in treatment. Xiong bi or chest obstruction is the classification in Chinese medicine. Coronary artery disease is the key feature within the mechanisms of the disease and the genetic determinants. Signs include low oxygen distribution, blood clots, arterial spasms and plaque deposits in the main and micro arteries of the heart. Health care providers should consider the types of angina, the genetic determinants and mechanisms, seasonal factors, and lifestyle choices when personalizing the treatment approach.

Types of Angina

The four types of angina pectoris include stable, unstable, variant and microvascular. Stable angina involves regular episodes of chest pain which is relieved by rest and medication. Unstable angina involves irregular unpredictable episodes of chest pain and microvascular features and it occurs primarily in women.

Mechanisms and Genetic Determinants

The mechanisms involve low oxygen distribution, blood clot formation, arterial spasms and plaque build-up in the microvessels. The genetic factors

Integrative Cardiovascular Chinese Medicine. DOI: http://dx.doi.org/10.1016/B978-0-12-420014-2.00010-7

involve problems with cell aging and lysis, plaque formation, methionine synthesis, and failure to control homocysteine levels.

Seasonal Factors

- Cold weather in winter contributes to deficiency of qi and yang, which causes stress during physical effort
- Hot weather in summer contributes to loss of qi and yin fluid, which raises body temperature and causes sweating

Lifestyle Choices

- Consider genetic factors that contribute to increased risk of cardiovascular disease
- Reduce the amount of fats in the diet
- Wean off cigarette smoking and tobacco use

Treatment success using integrative medicine focuses mostly on prevention and control during the stable angina stage.

Key Points

- Angina pectoris is classified under xiong bi.
- There are four types: stable, unstable, variant and microvascular.
- Stable angina pain happens during physical effort and is relieved by rest and medication.
- Unstable angina is not relieved by rest but is relieved by medication, usually sublingual nitroglycerine.
- Variant angina occurs at night and can be fatal.
- Microvascular angina occurs primarily in women and is not relieved by rest or medication.
- *DKN2A* and *DKN2B* are genes involved with plaque formation.
- *MTAP* is involved in methionine production and homocysteine levels.
- Myocardial and Na^+/K^+ pump failures increase H^+ production which lowers pH.
- Troponin is a cardiac marker which elevates within a few hours after cardiac injury and persists for up to 14 days.
- BNP/NT pro BNP is a cardiac marker, and levels indicate increased risk of infarction.
- CK-MB (creatine kinase) is a cardiac marker which elevates when there is damage to myocardial cells.
- Coronary angiography and CABG: determines the magnitude of coronary artery block. Next, healthy vessels from another area of the body are surgically used to bypass or replace the blocked section.

PATHOGENESIS

Chinese Medicine

Angina pectoris is classified under xiong bi chest obstruction and heartache, and involves various disharmonies between the heart, liver, kidney and spleen. The pathogenic factors are qi deficiency, blood stasis and stagnation and phlegm stasis. The complication to avoid is the fatal separation of yin and yang.

Western Medicine

a. Stable angina:
   - Pain is relieved by rest and medicine
   - Pain lasts for a few minutes and is predictable
   - Low oxygen distribution
   - Narrow arteries blocked with plaque, especially coronary arteries
   - Difficulty with physical effort
   - Emotional upset
   - Smoking
   - Seasonal weather extremes
b. Unstable angina:
   - Pain is not relieved by rest or medicine
   - Arterial blockage by plaque and blood clots
   - Pain occurs during rest or physical effort, and is unpredictable
c. Variant angina:
   - Cornonary artery spasm
   - Occurs during sleep at night
   - Relieved by medicine
d. Microvascular:
   - Pain is longer in duration and is not relieved by medicine or rest

DETERMINANTS AND MECHANISMS

Genetic Determinants

DKN2A and *DKN2B*. These genes are involved with regulating cellular aging and lysis, and the formation of plaque within the arteries.

MTAP. High levels of homocysteine are a risk factor for atherosclerosis. This gene is involved in the processing of cellular waste products into methionine which metabolizes excess homocysteine. Mutations are involved in failures in these processes.

Constitutional Factors

See Table 13.1.

TABLE 13.1

Constitution	General Incidence of Angina Pectoris
Teenage and young adult	Establishing of lifestyle habits: childhood obesity poor dietary choices and age of onset of smoking habit may set the health condition for middle age
Middle-age	Metabolic changes and adaptation to severe weather during summer and winter months impose additional demand during physical efforts
General geriatric	Blood pressure changes, cardiac conduction problems and excess blood lipoprotein disorders, may also cause plaque, fibrosis and blood clots which block arteries

Mechanisms

a. Biochemical changes:
   - Lactate, potassium ion accumulation at the cardiac nerve contribute to pain sensation

b. Contractility changes:
   - Myocardial and Na^+/K^+ pump failures increase H^+ production, which lowers pH
   - Papillary muscle dysfunction is due to dysfunction of the left ventricle
   - Myocardial contractility decreases with left ventricular failures

EXAMINATION OF THE PATIENT

Visual Observation

Perform a neurologic examination for stroke by noticing the patient's gait upon entering the room and bilateral grip strength.

Listening

- Assess memory, judgment and speech for neurological problems before or after treatment
- Emotions: vexation, moody and temperamental restlessness

Physical Examination

- Measurement of temperature, waist circumference, weight and height
- Evaluate the effects of blood pressure by viewing the retina, which can predict future patterns of cardiovascular disease

TABLE 13.2

Lab	Detections
C-reactive protein	Elevations detect inflammation presence in the body
CBC- complete blood count	Detects anemia through the activity of blood components such as red blood cells, white blood cells and platelets
Metabolic panel	Checks lipoproteins, blood sugar, acid/base, liver enzymes, kidney functioning and electrolytes
Troponin	Marker which elevates within a few hours after cardiac injury and persists for up to 14 days
BNP/NT pro BNP	Increased levels indicate increased risk of infarction
CK-MB (creatine kinase)	Marker which elevates when there is damage to myocardial cells

a. Blood pressure:
- This is taken in a sitting, standing or supine position, and readings indicate stages of hypertension

b. Pulse in TCM:
- Chinese pulse diagnosis:
  - Irregular = weak heart qi
  - Tight = pain and cold syndrome
  - Weak = deficiency
  - Wiry = qi stagnation
  - Thready = hyperactive kidney
  - Rolling = dampness, phlegm

Auscultation

Heart:

- First palpate the apical impulse for left ventricular enlargement or hypertrophy
- Listen for an early sign of hypertension, the fourth heart sound (S4) which indicates left atrium overwork
- Listen for underlying signs of heart disease, the third heart sound (S3) which indicates left ventricula malfunction

LABORATORY TESTS

Routine Tests

See Table 13.2.

Other Tests

a. Chest X-ray: determines heart enlargement.
b. Stress test: determines the magnitude of symptoms during stress.
   - a very rapid heart rate due to pulmonary causes or ischemia
   - normal activity, but develops shortness of breath with exertion
c. Coronary angiography and CABG: determines the magnitude of coronary artery block. A catheter is inserted through an artery in the arm leg or neck towards the coronary arteries. A contrast dye is injected, providing a view or the artery condition on X-ray. Healthy vessels from another area of the body are surgically used to bypass or replace the blocked section.
d. Transthoracic echocardiogram. This test identifies the condition, size and dimensions of the left and right atria, the left and right ventricles, valves, and the presence of thrombus.

CLINICAL TREATMENT

Western Medicine

Work up and Treatment

- Family history of hypertension, heart disease or other systemic diseases
- BMI >30
- Mild hyperlipidemia
- Smoker and alcohol drinker
- High stress relationships, lifestyle or occupation
- Increase physical activity
- Diet: irregular, high fat and sugar, low in fresh fruits and vegetables
- Stop smoking

Chinese Medicine

See Figure 13.1.

a. Clinical presentation: pain sensation in the chest area, palpitations, dyspnea, spontaneous sweating.
   Tongue = red
   Pulse = weak
   See Figure 13.2.

Qi Deficiency

This patient may have early symptoms due to emotional upset or other lifestyle factors. Here the lung qi may also be deficient due to random periods of holding in of the breath.

FIGURE 13.1

Clinical Pearl:

The patient may have either a deficiency of yin or yang with this condition. The yin deficiency may include restlessness and dizziness, the yang deficiency may involve episodes of syncope. This is the crucial time of re-balancing yin and yang as a prevention effort against complications seen in hyperactivity.

FIGURE 13.2

Cold Stagnation in the Chest

When cold air is breathed in the sensation may appear to be an abrupt attack in the chest area, followed by lingering symptoms which appear elsewhere in the body, especially in the legs.

FIGURE 13.3

Clinical Pearl:

This kind of patient may not have asthma, but will have similar symptoms during cold weather months.

FIGURE 13.4

b. Treatment:
- Therapeutic principle: tonify qi, nourish blood, balance yin and yang
- Medicine: pulse-activating powder
- Exercise: tai qi

c. Modifiable medicated diet: gruel with astralagus, and Chinese date for breakfast.
See Figure 13.3.

a. Clinical presentation: palpitations, chest oppression with shortness of breath.
Tongue = pale with greasy white coat
Pulse = tight
See Figure 13.4.

b. Treatment:
- Therapeutic principle: warm the yang, remove chest obstruction
- Medicine: gui zhi tang
- Exercise: tai qi

c. Modifiable medicated diet:
- Gruel with cinnamon and ginger for breakfast in the morning
- Thin soup with seaweed onions and ginger with lunch and dinner

See Figure 13.5.

a. Clinical presentation: palpitations, chest pain and oppression, irritability, vertigo, tinnitus, insomnia and excessive dreaming during sleep.

Tongue = no significance

Pulse = wiry

See Figure 13.6.

b. Treatment:
- Therapeutic principle: soothe liver qi, nourish yin, harmonize heart and kidneys, tranquilize the mind
- Medicine: tian ma gou teng yin
- Medicated diet: gruel or soup using astralagus and ginseng for breakfast and lunch for energy production during the day

Choose herbs from the formula as a base for a soup at dinner and take the formula in the evening before bedtime.

See Figure 13.7.

a. Clinical presentation: oppressed chest sensation, palpitations, precordial and hypochondrial area pain, heavy body sensation, shortness of breath and exhaustion.

Hyperactivity of Liver Yang

In this differentiation, the liver yang hyperactivity complicates heart and kidney yin deficiency.

FIGURE 13.5

Clinical Pearl:

For this patient, a variety of formulas can be used for liver yang hyperactivity. The patient may experience tiredness and exhaustion during the day, and insomnia at night. The purpose of treatment with this differentiation is to bring down the blood pressure by soothing the liver, balancing yin and yang and heart and kidney, all at the same time.
Relief of hyperactivity symptoms may happen immediately, but will not affect the other symptoms.

FIGURE 13.6

Qi Stagnation with Blood Stasis

Qi and blood are not flowing freely through the vessels, which is often connected to damp accumulation and phlegm stagnation.

FIGURE 13.7

Clinical Pearl:

Patience, skill and insight are needed for management of this patient. They may have complicated symptoms with alternations between pulse qualities which may confuse the direction in treatment.
Caution during treatment if the condition alternated between qi stagnation and dampness accumulation.
A possible clinically induced damp heat may occur if the practitioner is too hasty in tonifying qi.
Also in some patients, the condition may necessitate hospitalization. Depending on the protocols there, Chinese medicine treatment in this differentiation category may be included as a dietary regimen which may help nourish the patient during recovery.

FIGURE 13.8

Exhausted Yin and Yang

Yin and yang are unable to promote and restrict as the deficiency of both leads eventually to end stage separation.

FIGURE 13.9

Tongue = pale, purple, greasy white coating
Pulse = irregular, wiry, rolling

See Figure 13.8.

b. Treatment:

- Therapeutic principle: free the flow of qi, invigorate blood and promote circulation and drain dampness
- Medicine: xue fu zhu yu tang

c. Modifiable medicated diet: gruel with hawthorn, fleeceflower root, chinaberry, and Chinese date for breakfast.

Thin soup with seaweed onions and ginger with lunch and dinner.
See Figure 13.9.

Clinical Pearl:

Most patients in this differentiation category may have had more experience with hospitalization treatments. In situations when the patient survives and is referred for Chinese medicine, gradual recovery of yin and yang may be the first task. Close monitoring of the lab reports may help guide a favorable prognosis.

FIGURE 13.10

a. Clinical presentation: dyspnea, palpitations, irritability, coldness, spontaneous sweating.

Tongue = no significance

Pulse = very deep, irregular and rapid

See Figure 13.10.

b. Treatment:

- Therapeutic principle: nourish yin and blood, tonify yang and qi
- Modifiable medicine: xu ming decoction

c. Modifiable medicated diet:

- Gruel with hawthorn, fleeceflower root, chinaberry, and Chinese date for breakfast. Thin soup with seaweed onions and ginger with lunch and dinner

MEDICINES

A. Western Medicine Drugs

See Table 13.3.

B. Chinese Medicine Formulas

See Table 13.4.

SUMMARY IN THE CONTEXT OF PATIENT PERSPECTIVES FOR PERSONALIZING MEDICINE

Health practitioners treating patients with hypertension may wish to consider information from the patient narrative, especially when the patient describes the symptoms of physical impairment during regular activities.

Angina pectoris is a serious condition which progresses over time. The purpose of early treatment is to prevent or control symptoms and to guide in reversing lifestyle factors which contribute to disease progression.

Patients who benefit the most using Chinese medicine treatments alone are those who may have not experienced a myocardial infarction. Once myocardial tissue injury or necrosis has occurred, integrating Chinese and

TABLE 13.3

Medicine	Action
Nitroglycerine	Vasodilator
Beta blocker	Blocks the binding receptor on heart, kidney, arteries, smooth muscle cells that respond to epinephrine, the stress response
Calcium channel blocker	Antihypertensive that decreases blood pressure
ACE inhibitor	Vasodilator that lowers blood pressure
Antiplatelet	Inhibits formation of thrombus
Anticoagulant	Inhibits formation of thrombus
Aspirin	Antiplatelet drug

TABLE 13.4

Medicine	Action
Xu ming decoction	Clears heat and activates blood flow
Xue fu zhu yu tang	Moves blood, promotes circulation, stops pain
Tian ma gou teng yin	Calms liver, clears heat, tranquilizes the mind
Gui zhi tang	Relieves the exterior, harmonizes wei qi and ying qi
Pulse activating powder	Replenishes qi, nourishes and astringes yin
Dan shen dripping pills	Promotes blood circulation, removes stasis, relieves pain

Western medicine may be indicated at the hospital inpatient level where close monitoring can guide more favorable results during the course of treatment. On a base level in treatment, ren shen (ginseng) alone, or featured in formulas, has the marvellous ability to rescue a patient on the brink of poor prognosis, and may prevent the separation of yin and yang.

FURTHER READING

[1] Guoliang S, Zhaohui P. Observation of the curative effect on Alprostadil injection combined with she xiang bao xin wan on angina pectoris. China Med J 2012;:1673–7210.

[2] Broadbent HM, et al. Susceptibility to coronary artery disease and diabetes is encoded by distinct, tightly linked SNPs in the ANRIL locus on chromosome 9p. Hum Mol Genet 2008;17(6):806–14.

[3] Helgadottir A, et al. A common variant on chromosome 9p 21 affects the risk of myocardial infarction. Science 2007;316(5830):1491–3.

Chapter 14

Rheumatic Heart Disease

Chapter Objectives

After studying this chapter you should be able to:

1. Explain the integrative medicine perspective on rheumatic fever
2. Describe the mechanisms of the disease
3. Discuss the Jones criteria for diagnosing
4. Discuss the pattern for preventive treatment using integrative Chinese medicine
5. Summarize the context of personalizing medicine

OVERVIEW

This chapter discusses the aspects of rheumatic fever which may be relieved using integrative medicine and may give insight into developing methods for personalizing medicine. Rheumatic fever is an autoimmune inflammatory disease. This disease often originates in childhood as untreated or undertreated strep throat caused by the group A beta-hemolytic streptococcus. Within 20 days after infection, symptoms of rheumatic fever can present and weaken the heart. Certain factors such as genetics and lifestyle may cause this to recur in some patients. The immune responses play a role in defense and also in the autoimmunity process. In some cases recurring symptoms later in life contribute to chamber and valve problems.

The Jones criteria are often used in the diagnosis of rheumatic fever; they identify the presence of rheumatic fever according to specific clinical presentations and laboratory tests. Childhood occurrence is characterized by the presence of chorea.

In Chinese medicine, pathogenesis begins with a failure of ying qi and wei qi to protect the body's defenses, leading to lingering exogenous heat and damp evils.

Integrative Cardiovascular Chinese Medicine. DOI: http://dx.doi.org/10.1016/B978-0-12-420014-2.00011-9

Key Points

- Wei qi and ying qi are part of the immune system and function to protect the body.
- Rheumatic fever is an autoimmune inflammatory disease which is connected with type II hypersensitivity, humoral and cellular immune-mediated immune injury.
- M proteins are similar to the antigen located on myocardial cells and therefore such antibodies may attack cardiac myocytes.
- In autoimmune situations, molecular mimicry involves B cells that produce antibodies which may attack cardiac tissues.
- The Jones criteria are used to assist in diagnosing.
- Chorea is a condition in affected children with temporary poor motor skills, emotional changes and contracture of the fingers.
- Carditis can be heard through auscultation, and may mark congestive heart failure, chest pain, pericardial friction and effusion.
- Polyarthritis includes migratory joint pain.
- Inflammation leads to Aschoff bodies or Anitschkow cells which are collagen fibrosis or lesions as in affected tissues.
- Temperature: 101 °F+ (38.2 °C+).
- Limbs: joint pain, subcutaneous nodules around the joints, and, in some patients, a body rash called erythema marginatum.
- Anti-streptolysin O test is often used for detecting *Streptococci* presence, which may be responsible for glomerular nephritis and joint pain caused by rheumatic fever or rheumatoid arthritis.

PATHOGENESIS

Chinese Medicine

In Chinese medicine, rheumatic fever belongs to bi zheng (bee-jung) or arthralgia syndrome. The pathogenesis begins with a failure of ying qi and wei qi to protect the body's defenses. Exogenous wind, damp and heat evils combine and invade the body, linger for a long period by gathering within the muscles, joints, skin, the blocking meridians, and later causing stagnant movement of qi and blood. Over time, as the effects of the disease recur, the heart is severely affected. The lungs are responsible for wei qi. In addition they can fail in the function of dispersing and descending.

Western Medicine

Rheumatic fever is an autoimmune inflammatory disease which is connected with type II hypersensitivity, humoral and cellular immune-mediated immune injury. Responses can activate the complement system and cause an inflammatory reaction in various parts of the body. Pathogenesis is caused primarily by the group A beta-hemolytic streptococcus.

(a) The M protein. M protein serotypes M1, M3, M5, M6, M19, M24 of streptococcus cell walls may cause the immune system to produce antibodies against it. This protein may be similar to the antigen located on myocardial cells, and therefore such antibodies may attack cardiac myocytes.
(b) $CD4^{+}$ T cells. In autoimmune situations, T lymphocyte activation will then activate B cells that produce antibodies which may attack cardiac tissues.
(c) VCAM-1. This protein is connected with the involvement of $CD4^{+}$ T cells in inflammation and valvular fibrosis.

DETERMINANTS AND MECHANISMS

Genetic Determinants

HLA-DR7 (human leukocyte antigen HLA). This protein is a MHC class II molecule which has a role of increasing the possibility of autoimmune reactions in some patients. It may be connected to $CD4^{+}$ responses.

Constitutional Factors

See Table 14.1.

Mechanisms

The long-term effects of rheumatic fever present as fibrosis and necrosis of the mitral and aortic valves.

TABLE 14.1

Constitution	General Incidence of Rheumatic Fever
Teenage and young adult	This infection is seen frequently in children, especially in signs such as chorea. In young adults untreated or undertreated streptococcal infection can result in complications within two weeks to three years. In these patients the infection needs to be cleared. Wei qi needs to be raised as a preventative against rheumatic heart disease and autoimmune complications later in life
Middle-age	Various microbial infection may be due to latent recurrence or recent exposure. Here the patient may present with some cardiovascular complications as a result of inflammation or during periods of illness
General geriatric	This patient may have advanced symptoms such as rheumatoid arthritis, autoimmune diseases and heart failure

(a) Type II hypersensitivity. This reaction is also known as molecular mimicry; it contributes to rheumatic fever. The cell-mediated immunity reaction occurs during a *Streptococcus* infection in which B cells, which should react to the microbial infection, begin to react only against the valvular tissue, myocardial cell, muscle and joint tissues.
(b) Inflammation. Neutrophils have a short life span for good reasons. One is to prevent complicating diseases. In the presence of inflammation, engulfs microorganisms and then are quickly phagocytized by macrophages. However, continued inflammation bringing more neutrophils may cause them to begin attacking tissues as well, but may later contribute to the development of Aschoff bodies or Anitschkow cells which are collagen fibrosis or lesions as in affected tissues.

EXAMINATION OF THE PATIENT

The Jones criteria are often used in the diagnosis of rheumatic fever; they identify the presence of rheumatic fever according to specific clinical presentations and laboratory tests. For a diagnosis, a patient must satisfy two major criteria and one minor criterion:

See Table 14.2.

Observation

- Temperature: 101 °F+ (38.2 °C+)
- Limbs: joint pain, subcutaneous nodules around the joints, and, in some patients, a body rash called erythema marginatum.

TABLE 14.2

Major Criteria	Findings
Carditis	Congestive heart failure, chest pain, pericardial friction and effusion
Polyarthritis	Migratory pain in joints of hands and other areas
Erythema marginaturn	Circular rash pattern on the trunk and limbs
Chorea	Found in children with temporary poor motor skills, emotional changes and contracture of the fingers
Minor Criteria	**Findings**
EKG	Prolonged PR interval
Laboratory	Elevated ESR and CRP
Clinical assessment	Elevated body temperature or joint pain

Physical Examination

(a) Blood pressure:
- This is taken in a sitting, standing or supine position, and readings indicate stages of hypertension.

(b) Auscultation:
- First palpate the apical impulse for left ventricular enlargement or hypertrophy.
- Listen at the apex or left sternal border for a possible loud opening snap in the mitral valves unless there is calcification. This can be heard during the first heart sound (S1) and second heart sound (S2).
- Listen for underlying signs of heart disease, the third heart sound (S3) – which indicates left ventricular malfunction.
- Listen for a pericardial rub, indicating a possible carditis – which is an inflammatory condition.

Laboratory Tests

Routine tests:

(a) EKG.
See Figure 14.1.
A long PR interval is characteristic.

(b) Blood work: usually a throat culture is done to detect the presence and numbers of streptococci.

See Table 14.3.

CLINICAL TREATMENT

Western Medicine

(a) Clinical presentation: past or recent history of strep throat or scarlet fever, abdominal pain, chorea, shortness of breath, skin rash, subcutaneous nodules on the arms.

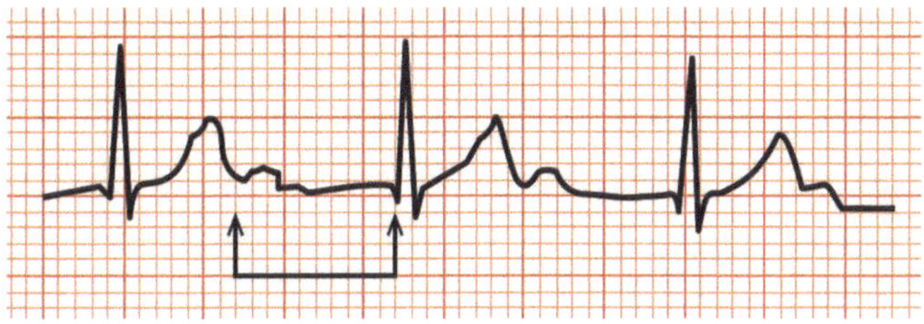

FIGURE 14.1

TABLE 14.3

Lab	Detections
C-reactive protein	Elevations detect inflammation presence in the body
ESR- erythrocyte sedimentation rate	Elevations detect inflammatory diseases
Anti-streptolysin O test	For detecting streptococcus, glomerular nephritis and determining whether joint pain is caused by rheumatic fever or rheumatoid arthritis
Complement assay	For detecting immunomediated diseases
CBC- complete blood count	Detects the activity of blood components such as red blood cells, white blood cells and platelets

Wind-Heat Damp

The condition transforms quickly from cold into heat. The dampness complicates the exterior from restoring defense. The heat and wind heat rises above to affect the throat, muscles and joints, and below to affect the urinary functioning and vaginal discharge.

FIGURE 14.2

(b) Work up and treatment:
- Family history of hypertension, heart disease or other systemic diseases complicated by chronic inflammation.

(c) Medicine:
- Aspirin
- Corticosteroids
- Antibiotics: penicillin, erythromycin, azithromycin.

Chinese Medicine

See Figure 14.2.

(a) Clinical presentation:
- coldness, edema, palpitations, insomnia, shoulder pain, joint pain, exhaustion

  Tongue = white coating

  Pulse = rapid, deep

  See Figure 14.3.

Clinical Pearl:

Most patients in this differentiation category may have had more experience with hospitalization treatments. In situations when the patient survives and is referred for Chinese medicine, gradual recovery of yin and yang may be the first task. Close monitoring of the lab reports may help guide a favorable prognosis.

FIGURE 14.3

Qi and Yang Deficiency

In this differentiation the breakdown of the body defense system includes: lung qi deficiency which damages wei qi, and spleen qi and yang deficiency which damages ying qi.

FIGURE 14.4

Clinical Pearl:

The qi deficiency may produce a rapid pulse, but in fact the patient is tired. The yang deficiency in this kind of patient is a yin pathogenic factor where the patient feels false wind cold-like symptoms which don't seem to be relieved by warm drinks, covering or warming remedies.

FIGURE 14.5

(b) Treatment:
- Therapeutic principle: tonify qi, raise yang
- Modifiable medicine:
  - sheng yang yi wei tang
  - ren shen tang

See Figure 14.4.

(a) Clinical presentation: fever, sore throat, malaise, chest discomfort, restless movements, thirst for cold drinks, desire for warmth. Joints which are red and painful to the touch.

Tongue = thick white coat with yellow center

Pulse = rapid

See Figure 14.5.

TABLE 14.4

Medicine	Action
Aspirin	Anti-inflammatory, analgesic, antipyretic
Corticosteroids	Anti-inflammatory
Penicillin	Antibacterial
Erythromycin	Antibacterial
Azithromycin	Antibacterial

TABLE 14.5

Medicine	Action
qing wen bai du yin	Treats severe toxic heat
pu ji xiao du yin	Treats severe toxic heat
long dan xie gan tang	Treats damp heat which affects the lower jiao
shen tong zh yu tang	Unblocks meridians and moves qi and blood

(b) Treatment:
- Modified medicine:
  - qing wen bai du yin
  - pu ji xiao du yin
  - long dan xie gan tang
  - shen tong zh yu tang

Medicines

A. Western medicine drugs
See Table 14.4.

B. Chinese medicine formulas

See Table 14.5.

SUMMARY IN THE CONTEXT OF PATIENT PERSPECTIVES FOR PERSONALIZING MEDICINE

Health practitioners treating patients with strep throat or toxic heat pathogens may find it necessary to provide liquid decoctions over a relatively long period of time (5−7 days) while at the same time cooperating with Western medicine pharmaceutical drugs. The flavor of antibiotic Chinese medicine antibiotic decoctions are bitter and strong, but effective. A suggestion may be to monitor the patient's condition weekly, introducing yang tonifying herbs until blood work shows that the body has eliminated the microbe. For certain patients, this approach may give an immediate feeling of balance, and blood pressure may become stabilized during treatment; the effects can be noticed by looking at the skin quality on the face. The patient may then be ready to organize diet and lifestyle modifications which can further personalize the treatment plan.

FURTHER READING

[1] Zhang H, Chen Y. Integrative treatment in 72 children with acute rheumatic fever. Chin Community Doctors 2011;13(262):114.

[2] Han Y, Sun J. Progress in diagnosis and treatment of rheumatic fever. J Clin Pediatr 2012;30(7):697−700.

[3] Wang J. Medicine. People's Medical Publishing House; 2005.

[4] Jiang H. Three cases of rheumatic fever in the same family. Clin J Med Off 2012;40(6): 1544−6.

[5] Yechen H. A study on tumor necrosis factor α-G308A polymorphism and its susceptibility to rheumatic fever. J Youjiang Med Univ Natl 2011;33(4):413−4.

[6] Li Y. Clinical analysis in 60 cases that with rheumatic fever. Chin J Mod Drug Appl 2012;6(21):40−1.

[7] TCM treatment technology of rheumatic fever. (2010). China News of Traditional Chinese Medicine, 5th ed.

[8] Liu X. Yixin huoxue evils treat rheumatic heart disease. Chin J Cancer Chin West Med 2011;1(1):111−3.

[9] Xie X, Zhou H, Dong R, et al. Diagnostic analysis of 70 cases of rheumatic fever that complicated by carditis. Chin J Clin 2011;5(2):523−5.

[10] Xu J, Li X. Experience on treatment of rheumatic fever according to syndrome differentiation in TCM. Xinjiang J Tradit Chin Med 2012;30(6):96−7.

[11] Liu D, Zhang S. Differentiation of traditional Chinese medicine combined with western medicine treatment of symptomatic medication in 52 cases with rheumatic heart disease. Clin J Tradit Chin Med 2011;23(11):964−5.

[12] Guo Y. The clinical efficacy of Chinese medicine treatment in rheumatic heart disease. Guide China Med 2011;9(16):125−6.

[13] Wang Q, Lu F. Clinical efficacy of dopamine furosemide nitroglycerin mixture treat rheumatic heart disease combination heart failure. For All Healthy 2013;7(3):17−8.

Chapter 15

Endocarditis

Chapter Objectives

After studying this chapter you should be able to:

1. Explain the disease complications
2. Explain the indications for treatment
3. Describe the role of Chinese medicine during the indication stage
4. Discuss treatment methods in Chinese medicine
5. Summarize the context of the patient role in personalizing medicine

OVERVIEW

This chapter describes endocarditis which is caused by factors discussed in Chapter 14 (rheumatic fever). Integrating Chinese medicine and Western medicine may be more useful during early stages before the development of the condition. There are two types of endocarditis: *acute endocarditis* occurs as a result of recent infection, while *subacute endocarditis* is a chronic condition. Patients with congenital heart defects may require additional monitoring. The heart valves are the main target for bacterial, fungal or viral replication. Vegetations form and encrust in oscillating clusters which frequently break off into the blood flow, travel through the cardiopulmonary system, and infect other areas of the body. In the brain, vegetations can cause meningitis.

Chinese medicine determines that the cause is a continuation of the factors of rheumatic fever, with a failure of ying qi and wei qi to protect the body. Exogenous wind, damp and heat evils transform into endogenous toxic evils which consume qi and blood and complicate disease within muscles, joints, skin, and the meridians, and threaten the life force of the patient.

The mechanisms involve fibrosis, and thrombosis accumulation provides the medium for bacterial growth and also causes ventricular septal defects and atrial septal defects; it also affects the endothelium within the vessels.

Integrative Cardiovascular Chinese Medicine. DOI: http://dx.doi.org/10.1016/B978-0-12-420014-2.00012-0

Genetic determinants are involved in high susceptibility to autoimmune and inflammatory complications such as in multiple sclerosis and systemic lupus, and renal diseases.

Special complications:

- Heart: murmur, heart failure, valvular damage
- Brain: emboli or meningitis
- Lungs: pneumonia, emboli due to right heart endocarditis
- Kidneys: glomerular nephritis due to a metastasized piece of vegetation
- Spleen: splenomegaly due to emboli

Nosocomial endocarditis: contraction due to infection of artificial pacemakers, implants and intravenous lines.

Key Points

- Exogenous wind, damp and heat evils transform into endogenous heat toxins.
- Organisms acquired in the community and health care setting, gastrointestinal or genitourinary problems.
- *HLA-DQA1*, *CFHR1* and *CFH3* are genetic determinants involved in predisposition to autoimmune diseases and nephritic syndrome.
- Rheumatic fever has the possibility of transforming into endocarditis later in life for some patients.
- Duke criteria are often used in the diagnosis of endocarditis; they determine whether the patient presents with (i) two major criteria, (ii) one major and three minor criteria, or (iii) five minor criteria.
- 'FROM JANE' is a mnemonic for endocarditis: **F**ever, **R**oth's spots, **O**sler's nodes, **M**urmur, **J**aneway lesions, **A**nemia, **N**ail hemorrhage, **E**mboli.
- Acute endocarditis may be relieved using integrative medicine, as the focus is to nourish yin, activate and move blood stasis, invigorate and tonify qi and warm yang.
- Chronic endocarditis often requires frequent hospitalization as the condition begins to cause heart failure, and Chinese medicine treatments may not be indicated.
- A positive Allen's test may indicate a circulation problem or Buerger's disease.

PATHOGENESIS

Chinese Medicine

In Chinese medicine, endocarditis is related to rheumatic fever where the pathogenesis begins with a failure of ying qi and wei qi to protect the body defenses. Exogenous wind, damp and heat evils transform into endogenous heat toxins which consume qi and blood and complicate disease within muscles, joints, skin, the gastrointestinal system and meridians, and threaten the life force of the patient. Over time, as the effects of the disease recur, the heart is severely affected. The lungs are responsible for wei qi. In addition, they can fail in the function of dispersing and descending – with edema and dyspnea as the result.

Treatment may concentrate on methods to assist with controlling and eliminating heat toxin and pathogenic factors, nourishing and invigorating blood, invigorating qi and warming yang.

Western Medicine

In patients who develop endocarditis, the left side of the heart is often affected. The mitral and aortic valves fail due to the development of vegetation: vegetations is the collection of organisms which encrust the valves, causing shape and function changes.

(a) Organisms acquired in the community and health care setting:
- HACEK organisms
- *Staphylococcus*
- *Streptococcus*
- Viridans group *Staphylococcus*
- *Candida albicans*
- *Histoplasma capsulatum*
- *Aspergillus*
- *Pseudomonas*

(b) Organisms acquired due to gastrointestinal or genitourinary problems:
- *Enterococcus*
- *Streptococcus bovis*
- *Clostridium septicum*

DETERMINANTS AND MECHANISMS

Genetic Determinants

(a) *HLA-DQA1*. This gene is associated with autoimmunity and idiopathic membranous nephropathy, which causes nephrotic syndrome and leads to end-stage renal disease.

(b) *CFHR1* and *CFH3*. These genes, which are part of the major histocompatibility complex (MHC), are involved in IgA nephropathy, which is connected to diseases with autoimmune and inflammatory complications such as multiple sclerosis and systemic lupus.

Constitutional Determinants

See Table 15.1.

TABLE 15.1

Constitution	General Incidence of Endocarditis
Teenage and young adult	Rheumatic fever is seen in children. In young adults untreated or undertreated streptococcal infection can result in complications within two weeks to three years. In these patients the infection needs to be cleared. Wei qi needs to be raised as a preventative against rheumatic heart disease, autoimmune diseases and development of endocarditis later in life
Middle-age	Various microbial infection may be due to latent recurrence or recent exposure. Here the patient may present with some resistant cardiovascular complications or occurrences connected with seasonal flu-like illnesses
General geriatric	This patient may have advanced symptoms which may be life threatening which require antibiotic and surgical therapy

Mechanisms

Damaged valves may be congenital deformities or may arise from surgical procedures or autoimmune disease, or as a common occurrence in the elderly. The valves of the right and left heart, and the endocardium, are the sites of development of endocarditis. Fibrosis and thrombi accumulation provide the medium for bacterial growth, and also cause ventricular septal defects and atrial septal defects; they also affect the endothelium within vessels.

EXAMINATION OF THE PATIENT

Like the Jones criteria for diagnosing rheumatic fever (mentioned in Chapter 14), the Duke criteria are often used in the diagnosis of endocarditis. The diagnostic scheme involves: (i) two major criteria, (ii) one major criterion and three minor criteria, or (iii) five minor criteria (see Table 15.2).

Visual Observation

(a) Perform a neurologic examination for stroke by noticing the patient's gait upon entering the room, and also bilateral grip strength.

(b) 'FROM JANE' – a mnemonic for endocarditis:
- Fever
- Roth's spots: spots seen on the retina during an eye examination
- Osler's nodes: painful lesions on the ends of the fingers

TABLE 15.2

Major Criteria	Findings
Typical micro-organisms These factors are dependent on: • Anti-phase I IgG antibody • Two positive cultures of blood samples drawn more than 12 hours apart	• Viridans-group, or • Streptococcus bovis, or • HACEK, or • Staphylococcus aureus, or • Enterococci
Echocardiogram evidence reporting:	• Oscillating mass on a heart valve or endocardium, or • Vegetation on prosthetic device, or • Valvular regurgitation
Minor Criteria	**Findings**
Immunological problems	Glomerulonephritis, or Osier's nodes
Fever	More than 101°F or 38°C
Clinical assessment	• Cardiac lesions • Recreational drug use with needles • Janeway lesions • Conjunctival hemorrhage • Arterialemboli • Positive blood culture outside of the major criteria

- Murmur: heart murmur heard on auscultation
- Janeway lesions: painless lesions on the palms and soles
- Anemia: blood deficiency
- Nail hemorrhage: brownish splinter marks on the nail bed
- Emboli

Physical Examination

Measurement of temperature, waist circumference, weight and height.

Evaluate the effects of blood pressure by viewing the retina, which can predict future patterns of cardiovascular disease.

(a) Blood pressure. This is taken in a sitting, standing or supine position, and readings indicate stages of hypertension.

(b) Allen's test:

1. Have the patient make a tight fist which forces the blood out of the hands.

2. Press on the arteries of both wrists to control the blood from circulating into the hands.
3. Have the patient open the hands as you release pressure on each wrist separately.

If the flow of blood into the hands is slow, it may indicate a circulation problem or Buerger's disease.

Auscultation

Heart:

- First palpate the apical impulse for left ventricular enlargement or hypertrophy.
- Listen at the apex or left sternal border for a possible loud opening snap in the mitral valves followed by a click, unless there is calcification. This can be heard during the first heart sound (S1) and second heart sound (S2).
- Listen for an underlying sign of heart disease: the third heart sound (S3), which indicates left ventricular malfunction.
- Listen for a pericardial rub, indicating a possible carditis – which is an inflammatory condition.

LABORATORY TESTS

Routine Tests

See Table 15.3.

Other Tests

(a) Transthoracic echocardiogram. This test identifies the condition, size and dimensions of the left and right atria, the left and right ventricles, valves, and the presence of thrombi.

TABLE 15.3

Lab	Detections
Anti-streptolysin O test	For detecting streptococcus, glomerular nephritis and determining whether joint pain is caused by rheumatic fever or rheumatoid arthritis
CBC- complete blood count	Detects the activity of blood components such as red blood cells, white blood cells and platelets
C-reactive protein	Elevations detect inflammation presence in the body

(b) 12-lead electrocardiogram. This test is used to identify certain patterns which provide evidence of left ventricular hypertrophy, prior myocardial infarctions and various other forms of arrhythmia.

CLINICAL TREATMENT

Western Medicine

Additional clinical presentation:

- Fever
- Malaise
- Flu-like symptoms: dyspnea, aching muscles and joints, night sweats, headaches
- Edema: abdomen, legs and feet
- Skin changes on fingers, palms and soles
- Nail splinter hemorrhages
- Nausea and weight loss
- Blood or mucus in the urine

Chinese Medicine

a. Clinical presentation: coldness, edema, palpitations, insomnia, shoulder pain, joint pain, exhaustion, fever, sore throat, malaise, chest discomfort, restless movements, thirst for cold drinks, desire for warmth. Joints which are red and painful to touch.
   Tongue = white coating with yellow in the center
   Pulse = rapid, deep
b. Treatment
   - Modified medicine:
     - qing wen bai du yin
     - pu ji xiao du yin
     - long dan xie gan tang
     - shen tong zh yu tang

MEDICINES

A Western Medicine Drugs

See Table 15.4.

B Chinese Medicine Formulas

See Table 15.5.

TABLE 15.4

Medicine	Action
Intravenous vancomycin	Treats *Enterococcus* infection
Intravenous ceftrioxene	Treats resistant bacterial infections
Penicillin	Treats *Enterococci, Gemella sp., Granulicatella sp.* And *Abiotrophia defectiva*
Nafcillin	Treats *Staphylococcus* aureus
Aminoglycosides	Antibiotics which treat aerobic gram negative bacterial infections
Intravenous benzyl penicillin	Treats *Streptococcus* bovis

TABLE 15.5

Medicine	Action
Sheng yang yi wei tang	Resolves exterior pathogenic factors, drains dampness, raises yang
Ren shen tang	Replenishes qi, nourishes yin, raises yang, treats palpitations and vertigo
Qing wen bai du yin	Clears toxic heat from blood, nourishes yin
Pu ji xiao du yin	Clears toxic heat and wind damp-heat, reduces swelling
Long dan xie gan tang	Clears excess liver and gall bladder damp heat
Shen tong zhu yu tang	Activates blood circulation and qi flow, dredges meridians and collaterals and relieves pain

SUMMARY IN THE CONTEXT OF PATIENT PERSPECTIVES FOR PERSONALIZING MEDICINE

As with rheumatic fever, health practitioners treating patients with acute endocarditis toxic heat pathogens may find it necessary to provide liquid decoctions over a longer period of time (12–30 days). The flavor of antibiotic Chinese medicine decoctions is bitter and strong, and cold but effective. A suggestion may be to monitor the patient's condition weekly, protecting the stomach from the cold herbs until blood work shows that the body has mostly eliminated the microbes. For certain patients, this medicine may be difficult to ingest because of the taste, but it can give the patient a feeling of balance, blood pressure may become stabilized during treatment, and the patient may revitalize. Finally, begin introducing qi- and yang-tonifying, blood-moving, and yin-nourishing herbs, while adding a few extra antimicrobial herbs to the decoction for an additional 6 months to 1 year. Some patients may need a decoction or extract containing only ren shen to help initial recovery; then the practitioner can begin to address the blood deficiency while monitoring the progress on CBC blood results. As early as possible during treatment, encourage the patient to organize diet and lifestyle modifications which can further personalize the treatment plan.

FURTHER READING

[1] Meng Ji. TCM syndrome differentiation and treatment of infective endocarditis. Chin J Basic Med Tradit Chin Med 2010;16(9).

[2] Nair R, Walker PD. Is IgA nephropathy the commonest primary glomerulopathy among young adults in the USA? Kidney Int 2006;69:1455–8.

[3] Xiaoxia A, Yunfang Z. Study on diagnostic criteria and methods for infective endocarditis. Chin J Pediatr 2008;23(10).

[4] Guo Gang W, Daoguo Z, An Hua Y. Clinical value of echocardiography in diagnosis of infective endocarditis. J Nosocomiol 2013;23(5).

[5] Hastings MC, et al. Galactose-deficient IgA1 in African Americans with IgA nephropathy: serum levels and heritability. Clin J Am Soc Nephrol 2010.

[6] Shengping H, et al. Analysis of 112 cases of infectious endocarditis. J South Med Univ 2012;33(2):279–81.

[7] Lin X, et al. Aberrant galactosylation of IgA1 is involved in the genetic susceptibility of Chinese patients with IgA nephropathy. Nephrolo Dial Transplant 2009;24:3372–5.

[8] Huan-ling W, Rui-yuan S. Aclinical analysis of 70 cases of infective endocarditis. Chin J Intern Med 2004;43(1).

[9] Gharavi AG, et al. IgA nephropathy, the most common cause of glomerulonephritis, is linked to 6q22-23. Nat Genet 2000;26:354–7.

[10] Austin III HA, Antonovych TT, MacKay K, Boumpas DT, Balow JE. Membranous nephropathy. Ann Intern Med 1992;116:672–82.

[11] Davila S, et al. Genome-wide association study identifies variants in the CFH region associated with host susceptibility to meningococcal disease. Nat Genet 2010;42:772–6.

[12] Hofstra JM, Wetzels JF. Alkylating agents in membranous nephropathy: efficacy proven beyond doubt. Nephrol Dial Transplant 2010;25:1760–6.

[13] Durack D, Lukes A, Bright D. New criteria for diagnosis of infective endocarditis: utilization of specific echocardiographic findings. Duke Endocarditis Service. Am J Med 1994;96(3):200–9.

[14] Yu, et al. Role of procalcitonin in the diagnosis of infective endocarditis: a meta-analysis. Am J Emerg Med 2013;31(6):935–41.

Chapter 16

Cardiogenic Shock

Chapter Objectives

After studying this chapter you should be able to:

1. Explain the integrative Chinese medicine treatment protocols for the survivor of cardiogenic shock
2. Describe the classification of shock in Chinese medicine
3. Describe the causes of shock
4. Summarize the context of the patient role in personalizing medicine

OVERVIEW

This chapter describes a clinical perspective of integrating Chinese medicine and Western medicine to recover a surviving patient who has experienced cardiogenic shock.

The explanation in Western medicine is that the patient's state is due to health conditions which cause an immediate lack of oxygen-rich blood circulating through the brain, heart and kidneys. Systolic blood pressure may be <80 mmHg for at least 30 minutes, cardiac index $</=2.2$ L/min m^2, pulmonary capillary pressure is $>/=15$ mmHg and left chamber end diastolic pressure is greater than 18 mmHg and right ventricular end diastolic pressure is greater than 10–15 mmHg.

In Chinese medicine, shock belongs to jue and tuo of both yin and yang, which means the complete exhaustion of yin and yang. Blood deficiency and depleted qi are factors to consider during treatment. In some cases it is necessary to rescue yang first, while in other cases it may be necessary to tonify qi and nourish yin first. In all cases syndrome differentiating is still important.

The combination of Chinese medicine and Western medicine treatment is for inpatient care in the hospital setting.

Integrative Cardiovascular Chinese Medicine. DOI: http://dx.doi.org/10.1016/B978-0-12-420014-2.00013-2

Key Points

- In Chinese medicine, shock belongs to jue and tuo of both yin and yang, which means the complete exhaustion of yin and yang.
- The most common causes of cardiogenic shock are myocarditis, endocarditis and arrhythmia leading to myocardial infarction.
- IL-6 and TNF-α contribute to reduced myocardial contractility and damage to blood vessel endothelium, especially in cases of sepsis.
- Right ventricle failure decreases output in both ventricles; left ventricle failure decreases output and hypoperfusion in the coronary arteries.
- Signs and symptoms of shock: sweating, pale skin, cool hands and feet, sudden rapid heartbeat, rapid breathing, confusion, loss of consciousness, weak pulse, decreased or no urine output.
- At the first sign of symptoms, call for emergency first-responder assistance.
- If the patient loses consciousness, press on DU 26 continually until the revival of the patient.
- Provide CPR assistance.
- A reading of <80 mmHg and pulse rate of </ = 70 or below for at least 30 minutes may also be evidence of shock.
- Emergency care in the private or outpatient clinic can include ren shen (ginseng), or sheng mai san which may help rescue yang, qi and yin as first-responder assistance is arriving.

PATHOGENESIS

Chinese Medicine

In Chinese medicine, shock belongs to jue and tuo of both yin and yang, which means the complete exhaustion of yin and yang. Other patterns include qi and phlegm obstruction, and the dissention of the normal five-element cycle leading to complete disharmony between the heart, spleen, lungs, kidneys and liver. In some cases it is necessary to rescue yang first, while in other cases it may be necessary to tonify qi and nourish yin first. In all cases syndrome differentiating is still important when treating the surviving patient.

Western Medicine

The most common causes of cardiogenic shock are myocarditis, endocarditis and arrhythmia leading to myocardial infarction. Other immediate concerns include the brain regulation of all other functions in the body. Failure of the kidneys may include metabolic problems. Liver failure may involve clotting and hemorrhage.

MECHANISMS

(a) Output. Peripheral resistance is complicated by ischemic conditions with the effect of hypoperfusion in the vessels, while at the same time the renin–angiotensin system may be involved in changes to blood volume, increases after load which leads to decreased cardiac output.

(b) Inflammation. IL-6 and TNF-α contribute to reduced myocardial contractility and damage to blood vessel endothelium, especially in cases of sepsis.

(c) Right ventricle dysfunction. Right ventricle failure decreases output in both ventricles, as the intraventricular septum shifts position.

(d) Left ventricle dysfunction. Left ventricle failure decreases output and hypoperfusion in the coronary arteries.

EXAMINATION OF THE PATIENT

(a) Signs and symptoms of shock in the clinical setting:
- Sweating
- Pale skin
- Cool hands and feet
- Sudden rapid heartbeat
- Rapid breathing
- Confusion
- Loss of consciousness
- Weak pulse
- Decreased or no urine output

(b) Responding to symptoms of shock in the private clinical setting. Shock could occur during acupuncture, ordinary medical office procedures, and other modalities in alternative medicine. At the first sign of symptoms, call for emergency first-responder assistance. Next, if the patient is conscious, try to administer 2 or 3 vials of concentrated ren shen (ginseng), or sheng mai san which may help rescue yang, qi and yin within 10–30 minutes. If the patient is still conscious, prepare to check the blood pressure. A reading of <80 mmHg and pulse rate of </= 70 or below for at least 30 minutes may also be evidence of shock. As the symptoms progress, place the patient in a comfortable position – or if the patient collapses to the floor, don't move them. If the patient loses consciousness, press on DU 26 continually until the revival of the patient. Provide cardiopulmonary resuscitation (CPR)

CLINICAL TREATMENT

Western Medicine and Chinese Medicine Hospital Inpatient Intravenous Care

See Figure 16.1.

For a patient who satisfies the following criteria, provide treatment two times per day for 1–2 weeks:

1. conscious and aware
2. pulse rate less than 100 beats/minute

L-Carnitine with Dopamine Injection

This medicine helps to improve recovery from myocardial ischemia and hypoxia.

FIGURE 16.1

Huang Qi Injection

Huangqi saponins provide protection against free radicals during myocardial ischemia and reperfusion. The medicine acts as an anti-coagulant and anti-thrombotic.

FIGURE 16.2

Kang Xiu Ning Decoction

This medicine acts as an anti-coagulant and anti-thrombotic, it also increases blood pressure and heart rate.

FIGURE 16.3

Shenqi Fu Zheng Injection

Ren shen and huang qi saponins can improve blood viscosity, blood pressure and heart rate.

FIGURE 16.4

3. pale complexion or signs of cyanosis
4. no urine or output less than 17 ml/hour
5. systolic pressure less than 80 mmHg

See Figures 16.2–16.4.

For a patient who satisfies the following criteria, provide treatment once per day for 1–2 weeks:

1. conscious and aware
2. pulse rate more than 150 beats/minute

Shenfu Injection

This medicine effectively increases the blood pressure, pulse pressure, expansion of coronary artery and reduces heart rate.

FIGURE 16.5

3. pale complexion
4. blood pressure more than 140/90 mmHg

See Figure 16.5.

SUMMARY IN THE CONTEXT OF PATIENT PERSPECTIVES FOR PERSONALIZING MEDICINE

Commonly, fatalities in cardiogenic shock range between 50% and higher based on constitution and age. If the hospital or clinical setting allows for elections between Western medicine alone or combination therapy of Chinese and Western medicine, then injection treatment with Chinese medicine can provide favorable results.

In the case of combination therapy, patients can be weaned off Western medicine treatment. The patient's condition can be stabilized while introducing Chinese medicine decoctions for preparing for patient discharge back to regular life. As early as possible during treatment, encourage the patient to organize diet and lifestyle modifications to address all disease conditions which contribute to cardiovascular disease.

FURTHER READING

[1] Gu M. Progress in mechanism and treatment of cardiogenic shock. Int J Cardiovasc Dis 2009;36(4):209–11.

[2] Tian Z. Analysis of salvage treatment of patients with cardiogenic shock. Appl Mod Chin Med 2011;5(6):144–5.

[3] Li C, Wang X. Mechanical circulation assist device in the treatment of cardiogenic shock in progress. J Prog Cardiovasc Dis 2012;33(1):36–8.

[4] Ma H, Liu J. With combination of traditional Chinese and Western medicine in treating acute myocardial heart source shock. Jilin J Tradit Chin Med 2002;22(3):29.

[5] Mi Z, Li Y, Zhang D. Observation of Huangqi injection in the treatment of acute myocardial infarction complicated by cardiogenic shock. JETCM 2009;18(10):1621, 1647.

[6] Ning B, et al. Clinical study of hemofiltration in treatment of cardiogenic shock. J Gen Hosp Air Force 2010;26(1):38–9.

[7] Chen Z. Shenfu injection on clinical observation on the treatment of cardiogenic shock. China J Clin Ration Drug Use 2012;5(9B):46–7.

[8] Gao Y, Yang J. Anti-xiuning decoction combined with Western medicine in treatment of 30 cases of cardiogenic shock. Liaoning J Tradit Chin Med 2011;38(5):869–70.

[9] Zheng L. Shenqi fuzheng injection in the treatment of cardiogenic shock effect and nursing care. Strait Pharm J 2012;24(11):143–4.

[10] Yi Y, Tian Y. Clinical observation of integrative medicine treatment of cardiogenic shock. Liaoning J Tradit Chin Med 2007;34(4):491–2.

[11] Tian. Z. Analysis of rescue processing in cardiogenic patients. Chin J Mod Drug Appl 2011;5(6):144–5.

[12] Huang Z. Advance in the therapy of cardiogenic shock. Chin J New Drugs Clin Remedies 2009;6:421–5.

[13] Mi Z, Li Y, Zhang D. Observation of Huangqi injection in treatment of acute myocardial infarction complicated by cardiogenic shock. J Emerg Tradit Chin Med 2009;10:033.

[14] Chen Z. Observation of Shenfu injection in treatment of acute myocardial infarction complicated by cardiogenic shock. Chin J Clin Ration Drug Use 2012;5(26):46–7.

[15] Xiang S. Left-carnitine combined with dopamine in treatment of acute myocardial infarction complicated by cardiogenic shock. China Mod Med 2010;17(007):53–4.

[16] Gao Y, Yang. J. Clinical observation of Kang Xiu Ning decoction combined with Western medicine with cardiogenic shock of 30 cases. Liaoning J Tradit Chin Med 2011; 38(5):869–70.

[17] Ning B, Zhang C, Ma Y, et al. Clinical study of continuous hemofiltration in treatment of cardiogenic shock. J Gen Hosp Air Force 2010;26(001):38–9.

[18] Yi Y, Tian Y. Clinical observation of integrative medicine treatment of cardiogenic shock. Liaoning J Tradit Chin Med 2007;34(4):491–2.

[19] Ma H, Liu J. Treatment of cardiogenic shock due to acute myocardial infarction by combined method of Chinese with Western medicine. Jilin J Tradit Chin Med 2002;3:031.

[20] Zheng L. Clinical observation and nursing of shenqifuzheng injection in treatment of cardiogenic shock. Strait Pharm J 2012;24(11):143–4.

[21] Sarswat N, Hollenberg SM. Cardiogenic shock. Hosp Pract 2010;38(1):74–83.

PART V

Vascular Diseases

Chapter 17

Thromboangiitis Obliterans

Chapter Objectives

After studying this chapter you should be able to:

1. Explain the integrative Chinese medicine treatment options for vasculitis
2. Describe the periods of disease progression
3. Describe the main features in diagnosis
4. Summarize the context of the patient role in personalizing medicine

OVERVIEW

This chapter describes a clinical perspective of thromboangiitis obliterans (TAO).

In Chinese medicine, TAO is also called tuo ju. In Western medicine it is also known as thromboangiitis obliterans is also known as Buerger's disease and vasculitis. The condition is thought to occur as a result of tobacco use. The disease is characterized by inflammation and thrombosis of small and medium arteries of the legs and feet – which recurs and progresses in stages. During the ischemic period, the veins of the lower leg are inflamed, the skin is ulcerated, and there is gangrene with intermittent claudication. During the nutritional disorder period there is persistent pain and muscle atrophy. During the necrosis period, the pain becomes worse as infection leads to the development of ulcers and gangrene.

As in Western medicine, treatments using Chinese medicine should be seen as a palliative care option which mostly addresses symptoms.

Key Points

- TAO is also known as gangrene and tuo ju.
- This disease is found in males aged 20–40 who are heavy smokers.
- Higher levels of estrogen and nitric oxide metabolites have been found in males with TAO.

Integrative Cardiovascular Chinese Medicine. DOI: http://dx.doi.org/10.1016/B978-0-12-420014-2.00014-4

- Check the state of the dorsal pedis pulse.
- Prognosis is often poor.
- The cardiovascular complications include arrhythmia and myocardial ischemia.

PATHOGENESIS

Chinese Medicine

In Chinese medicine, TAO is also known as vasculitis and belongs to the category of gangrene. The pathogenic factors include qi and blood deficiency, blood stasis, cold accumulation and toxic heat affection. There are five characteristic of tuo ju:

- congenital deficiencies
- weak kidney qi, yin and yang
- cold and dampness
- abnormal diet
- poor psycho-emotional response

Treatment courses are often effective; however, relapse and amputation represent a realistic prognosis.

Western Medicine

This disease is found in males aged 20–40 who are heavy smokers. The vessels of the lower legs become inflamed and ischemic, leading to claudication and necrosis. Prognosis is often poor, resulting in amputation. The cardiovascular complications include arrhythmia and myocardial ischemia. There are three periods of disease progression:

1 Ischemic Period

- pale complexion, feelings of coldness and fatigue
- veins of the lower leg are red, painful and cord-like
- possible ulcers, gangrene and secondary infection
- mild, intermittent claudication

2 Nutritional Disorder Period

- persistent pain in the rest and more serious at night
- fever and flushed complexion
- toe nail thickening and deformation
- muscle atrophy and thin appearance

3 Necrosis Period

- pain becomes worse

- infection becomes more serious as the affected limbs develop ulcers and dry gangrene and may appear black
- dry gangrene can become wet gangrene
- possible systemic poisoning becomes more life-threatening

DETERMINANTS AND MECHANISMS

Genetic Determinants

(a) Human leukocyte antigen (HLA). This includes the locus of immune system genes related to the major histocompatibility complex (MHC). These genes are divided into three classes:

Class I are related to digestion.

Class II are related to antigen formation.

Class III are related to the complement system or the innate system which is congenital and directs the immune activity over a person's lifetime. They play a role in diseases of autoimmunity.

(b) *MS4A1* gene. Deficiencies of both CD3 and CD20 are related to complications associated with T and B lymphocytes. T lymphocyte mediation leads to autoimmunity. Humoral immunity hyperfunction by B lymphocytes leads to vascular injury and thrombus formation.

Mechanisms

Hormonal problems:

- Higher levels of estrogen and nitric oxide metabolites have been found in males with TAO
- Some researchers believe that high levels of estrogen can cause vascular endothelial injury, which can cause thrombosis

EXAMINATION OF THE PATIENT

Visual Observation

Perform a neurologic examination for stroke by noticing the patient's gait upon entering the room and the bilateral grip strength.

- Assess memory, judgment and speech for neurologic problems before or after treatment with any medicine given for treatment, as claudication is a prominent complication
- Head: distended feeling, insomnia and confusion
- Ears: dizziness, tinnitus
- Mouth: bitter taste, dryness and salivation with nausea
- Emotions: vexation, moody and temperamental restlessness
- Limbs: pain, numbness and tingling in legs and feet

Physical Examination

(a) Blood pressure. This is taken in a sitting, standing or supine position, and readings indicate stages of hypertension.
(b) Pulse of the lower limb:
- Dorsal pedis pulse = weak or absent

Auscultation

Heart:

- First palpate the apical impulse for left ventricular enlargement or hypertrophy
- Listen for an early sign of hypertension, the fourth heart sound (S4) which indicates left atrium overwork
- Listen for an underlying sign of heart disease, the third heart sound (S3) which indicates left ventricular function

LABORATORY TESTS

Routine Tests

See Table 17.1.

CLINICAL TREATMENT

Western Medicine

(a) Clinical presentation. Claudication, lower leg with inflamed vessels, leg ulcers and gangrene. Pain and numbness in feet and toes. Depression and anxiety.
(b) Work up and treatment:
- Age: male 20–40
- Tobacco use, especially smoking

TABLE 17.1

Lab	Detections
CBC-complete blood count	Detects the activity of blood components such as red blood cells, white blood cells and platelets
C-reactive protein	Elevations detect inflammation presence in the body
Procalcitonin	Detects the bacterial and viral microbes responsible for sepsis, meningitis and pneumonia

(c) Treatment:
- Smoking cessation
- 100% hyperbaric oxygen
- Prostaglandins
- Corticosteroids
- Streptokinase

Bypass surgery: ablation of diseased vascular sections due to poor perfusion.

Debridement: removal of dead/necrotic tissue in ulcers.

Amputation: removal of a necrotic limb.

Chinese Medicine

1 Ischemic Period

See Figure 17.1.

Pulse: dorsal pulse is weak.

Treatment:

- Modifable medicine:
  - Tao hong si wu tang 3× daily with warm water
  - Si miao yong an tang 3× daily with warm water
  - Wu wei xiao du yin 3× daily with warm water

2 Nutritional Disorder Period

See Figure 17.2.

Pulse: dorsal is weak and deep.

Cold Accumulation with Qi Deficiency

Pale complexion, feelings of coldness and tiredness, intermittent claudication, pain in the lower leg with includes inflamed vessels, infection and formation of ulcers.

FIGURE 17.1

Cold Accumulation with Blood Stasis

Cold sensation of the skin on the legs with pain and muscle atrophy, intermittent claudication, toe nails begin to thicken and pain is worse at night.

FIGURE 17.2

Heat Toxin Type

Intermittent claudication, foot heat and swelling with ulcers and pus filled, foul smelling necrosis.

FIGURE 17.3

TABLE 17.2

Medicine	Action
Hyperbaric oxygen therapy	Vasodilation
Protaglandins	Vasodilation
Streptokinase	Antithrombosis
Corticosteroids	Anti-inflammatory

Treatment:

- Medicine:
  - Tao hong si wu tang 3× daily with warm water
  - Si miao yong an tang 3× daily with warm water
  - Wu wei xiao du yin 3× daily with warm water
  - Yang he tang 3× daily with warm water

3 Necrosis Period

See Figure 17.3.
Pulse: dorsal pulse cannot be felt.

Treatment:

- Therapeutic principle: clear heat, remove toxins, promote blood circulation
- Medicine:
  - tao hong si wu tang 3× daily with warm water
  - xian fang huo ming yin 3× daily with warm water
  - Tou nong san 3× daily with warm water

MEDICINES

A Western Medicine Drugs

See Table 17.2.

TABLE 17.3

Medicine	Action
Tao hong si wu tang	Breaks blood stasis, moves blood
Si miao yong an tang	Treats gangrene
Wu wei xiao du yin	Clears toxic heat and toxins
Xian fang huo ming yin	Clears toxic heat and toxins
Tou nong san	Removes toxins, replenishes qi and blood
Yang he tang	Treats gangrene

B Chinese Medicine Formulas

See Table 17.3.

SUMMARY IN THE CONTEXT OF PATIENT PERSPECTIVES FOR PERSONALIZING MEDICINE

Health practitioners using Chinese medicine to treat patients with vasculitis may find that it is a cost-effective method with results possibly at the same level as those in Western medicine. Treatment shows results and may delay amputation, but it is a slow process. Also, the effects of the herbs may vary from patient to patient.

When integrating Chinese medicine and Western medicine, it may be helpful for the patient to take pharmaceutical medications which can quickly reduce pain, oral Chinese medicine, which can address the movement of blood, the heat toxins and the purulent lesions, and then use the same decoctions as a wash or made into an ointment.

Another matter to address is the psychologic state of the patient due to factors in the condition and possibly the poor prognosis. The patient narrative will help health practitioners to realize what the patient is going through. Through syndrome differentiation, formulas can be modified to assist with this as well. Giving up tobacco use is the most important thing the patient can do during treatment, and also for cardiovascular reasons.

FURTHER READING

[1] Jin X, Wang M, Wu X, Zhang J. The investigation of the relationship between smoking and rat thromboangiitis obliterans [J]. Shandong Med J 2011;06:29–30.

[2] Guo Y, Dai Y. Immunopathology investigation of vascular wall of patients with thromboangiitis obliterans [J]. J Chongqing Med Univ 2009;12(1683):1685.

[3] Zhang J, Zhang S. The investigation of the role of estrogen in thromboangiitis obliterans [J]. J Shanxi Med Coll Contin Educ 2006;03:30–2.
[4] Wang Z. Clinical research on thromboangiitis obliterans treated by the optimized protocol of integrated TCM and WM [D]. Heilongjiang Univ Chin Med. 2012.
[5] Chen YW, Iwai T, Umeda M, et al. Elevated IgG titers to periodontal pathogens related to Buerger disease. Int J Cardiol 2007;122(1):79–81.
[6] Olin JW. Thromboangiitis obliterans (Buerger's disease). N Engl J Med 2000;343 (12):864–9.
[7] Cao Z, Xu H, Dong X, Qiao J. Changes of cardiac function in patients with thromboagiitis obliterans [J]. Shandong Med J 1981;01:12–3.
[8] Tang Y. Emotional nursing of thromboangiitis obliterans patients [J]. China Med Her 2009;11:89–90.
[9] Li G, Yang Y, Yang G, et al. Clinical application of autologous peripheral blood stem cells transplantation in treatment for thromboangiitis obliterans [J]. Chin J Bases Clin Gen Surg 2012;11:1187–90.
[10] Wang M, Wang F, Ji D, Li C. Interventional treatment of thromboangiitis obliterans of lower extremities [J]. J Interv Radiol 2012;10:850–4.
[11] Huo G. Clinical observation of Danhong injection combined with urokinase in the treatment of thromboangiitis obliterans. Chin J Clin Ration Drug Use 2013;March 6(3A).
[12] Li Z. Chinese medicine treatment of thromboangiitis obliterans [J]. Chin J Convalescent Med 2012;08:719–20.
[13] Li X. Zheng Xuejun in the treatment of thromboangiitis obliterans experience. Shanxi J TCM 2013;Jan 29(1).
[14] Ohta T, Ishioashi H, Hosaka M, et al. Clinical and social consequences of Buerger's disease [J]. J Vasc Surg 2004;39(1):176–80.
[15] H. Zhang, M. Li. (2008) Treatment of thromboangiitis obliterans in lower extremity [A]. Professional Committee of peripheral vascular diseases in association with Chinese and Western medicine in Zhejiang Province. Zhejiang Province in 2008 annual meeting of the assembly of vascular surgery [C]. Professional Committee of peripheral vascular diseases in association with Chinese and Western medicine in Zhejiang Province: 6.
[16] Miao Z, Chen X, Chen J. Therapeutic effect of Mailuoning injection on rat thromboangitis obliterans model[J]. Chin Med Mater 2008;06:880–2.
[17] He J, Yu W. Clinical observation on the treatment of 40 cases of thromboangiitis obliterans with DangguiHuoxue tablets [J]. Guiding J TCM 2007;01:34–35 + 43.
[18] Lin X. Vasculitis drugs of choice [J]. Strait Pharm J 2013;01:233–4.
[19] Wang X. Maixuekang capsule with *Erigeron breviscapus* and topical application of Qufushengji ointment in the treatment of thromboangiitis obliterans[J]. TCM Shaanxi 2007;02:161–2.

Chapter 18

Takayasu Arteritis

Chapter Objectives

After studying this chapter you should be able to:

1. Explain how integrative medicine is used to treat the disease
2. Describe phases of the disease from the perspectives of both Chinese medicine and Western medicine
3. Describe the criteria used for diagnosis
4. Discuss treatment methods in Chinese medicine and Western medicine
5. Summarize the context of personalizing medicine

OVERVIEW

This chapter discusses the etiology and progression of Takayasu arteritis, and both the Chinese medicine and Western medicine perspectives on treatment.

Takayasu arteritis is also called 'pulseless disease', and primarily affects females between the ages of 20 and 40. There are six types, which affect various major arteries, and which can lead to stenosis, and they are associated with kidney and heart failure. The mechanisms of the disease involve heredity, estrogen excess and infection. The genetic determinant involves interleukin-12B, which is connected to the autoimmunity aspect.

In Chinese medicine, Takayasu arteritis is classified under vessel bi syndrome and is related to complications of the disharmony between the liver, spleen and kidney. The condition is differentiated as toxic phlegm heat during the active phases of the disease and deficiency of qi and blood stasis which can transform into yang deficiency with yin excess during remission in the stable phase.

Treatment using Chinese medicine can work with Western medicine to relieve physical symptoms such as pain, weakness and emotional disturbances.

Integrative Cardiovascular Chinese Medicine. DOI: http://dx.doi.org/10.1016/B978-0-12-420014-2.00015-6

Key Points

- Takayasu arteritis is considered to be an autoimmune disorder which primarily affects the aorta of females between the ages of 20 and 40.
- It is a chronic inflammatory disease which affects large arteries and is characterized by the absence of a pulse, thus it is also called the 'pulseless disease'.
- In Chinese medicine, the disease is divided into active and stable phases.
- There are six types which affect sections of the aorta, abdominal and renal arteries.
- Interleukin and HLB are genetic determinants involved in the inflammatory autoimmune disease aspect of Takayasu arteritis.
- The disease mechanisms include heredity, estrogen excess and infection.
- Three factors from the 1990 American College of Rheumatology criteria can be used for diagnosis.
- Two major criteria and one minor criterion from the 1986 Ishikawa criteria can be used for diagnosis.

PATHOGENESIS

Chinese Medicine

In Chinese medicine, Takayasu arteritis is classified under vessel bi. It is considered an autoimmune disorder which primarily affects the aorta of females, and may be connected with viral and bacterial infections linked to rheumatic fever.

The disease is divided into active and stable phases:

- Active phase: pathogenic qi factors of toxic phlegm heat
- Stable phase: deficient qi and blood stasis which later transforms into deficient yang and excessive yin and deficiency of yin and yang

Western Medicine

Takayasu arteritis is a chronic inflammatory disease which affects large arteries and is characterized by the absence of a pulse; it is thus also called 'pulseless disease'. It primarily affects women between the ages of 20 and 50, with a series of vascular changes which occur in stages. These vascular changes lead to aortic, abdominal and renal artery stenosis. Cardiopulmonary failures may occur, featuring right heart failure. Some types of disease include cardiomyopathy, and some patients may develop pulmonary complications such as pneumonia and interstitial pulmonary fibrosis.

There are six types of Takayasu arteritis:

- Type I: This type affects the branches of the aortic arch
- Type IIa: This type affects the ascending aorta, aortic arch and branches

- Type IIb: This type affects the ascending aorta, aortic arch, branches and descending aorta
- Type III & IV: These types affect the descending aorta, abdominal and renal arteries
- Type V: This type affects the aorta and branches

DETERMINANTS AND MECHANISMS

Genetic Determinants

(a) There may be several risk factors within and outside the HLA-B and HLA-D loci: HLA-DQB1/HLA-DRB1, FCGR2A/FCGR3A, and PSMG1. The HLA-D locus is a class II antigen that is produced by B-lymphocytes.
(b) Interleukin-12B (IL-12B) and IL-23 are involved in inflammatory autoimmune disease.

Mechanisms

The factors involved in the disease etiology include heredity, estrogen excess and infection.

(a) Heredity. In Asian females and others with Asian ancestry, the condition may be connected to a predisposition to autoimmunity.
(b) Estrogen. In some patients, high levels of estrogen may affect the production and release of nitric oxide. This may also affect endothelial cell responses to injury, inflammation and repair.
(c) Infection. The disease may be connected to viral and bacterial infection, such as infection by species of *Streptococcus* or *Mycobacterium*.

EXAMINATION OF THE PATIENT

A positive diagnosis can be made on the basis of meeting three of the criteria shown in Table 18.1.

A positive diagnosis can be made on the basis of meeting two major criteria and one minor criterion, as shown in Figure 18.1.

PHYSICAL EXAMINATION

(a) Blood pressure. This is taken in a sitting, standing or supine position, and readings indicate possible hypertension or hypotension.
(b) Pulse in TCM:
    - Chinese pulse diagnosis:
        Rapid = true heat, yin deficiency heat, infection
        Forceful = compensation due to deficiency
        Weak = deficiency

TABLE 18.1

	1990 American College of Rheumatology criteria
Age of onset	<40 years old
Claudication	Development and worsening of fatigue and discomfort in muscles of 1 or more extremity while in use, especially the upper extremities
Brachial artery pulse	Decreased pulsation of 1 or both brachial arteries
BP difference	Difference of >10 mm Hg in systolic blood pressure between arms
Bruit	Bruit audible on auscultation over 1 or both subclavian arteries or abdominal aorta
Arteriogram abnormality	Arteriographic narrowing or occlusion of the aorta, primary branches, or large arteries in the proximal upper or lower extremities. This is not due to atherosclerosis or fibro-muscular dysplasia

1986 Ishikawa criteria	
Age of onset	< 40 years of age with at least 1 month of symptom duration.
Major criteria	
Angiographic findings include lesions in the left and right mid sub-clavian artery. The most severe stenosis or occlusion presents in the mid portion of the artery. The location is from a 1 cm point which is proximal to the left and right of the vertebral artery orifices to a 3-cm distal point to the orifice.	
Minor criteria	
Angiographic and echocardiographic findings include annulo-aortic ectasia or aortic regurgitation. Also there may be lesions found in the pulmonary artery, left mid–common carotid, distal brachio-cephalic trunk, descending aorta, or abdominal aorta.	

FIGURE 18.1

Wiry = hyperactive liver condition
Thready = hyperactive kidney

Auscultation

Listen over the subclavian, aortic and abdominal arteries for bruit.

LABORATORY TESTS

Routine Tests

See Table 18.2.

TABLE 18.2

Lab	Detections
C-reactive protein	This test is a direct marker for inflammation during the active disease process
ESR- erythrocyte sedimentation rate	Elevations detect inflammation but may be non-specific
CBC- complete blood count	Detects the activity of blood components such as white blood cells, red blood cells and platelets
Anti-streptolysin O test	For detecting streptococcus evidence which may determine rheumatic fever or rheumatoid arthritis as the cause of joint pain

TABLE 18.3

Lab	Detections
Computed tomography angiography (CTA)	Detects any narrowing, occlusion, aneurysm or wall thickening of vessels
Doppler ultrasound	Detects arterial occlusion, narrowing and dilation in the aorta and elsewhere in the body such as in the mesenteric and renal arteries
Magnetic resonance imaging angiography (MRA)	May detect early evidence of active disease state such as edema, wall thickening in inflammation

Other Tests

See Table 18.3.

CLINICAL TREATMENT

Western Medicine

(a) Clinical presentation
In the early phase:
- fatigue
- fever with night sweats
- cough
- pleural effusion
- weight loss

- abdominal pain with vomiting
- arthralgia
- elevated erythrocyte sedimentation rate
- high plasma protein

  In the pre-pulseless phase:
- during this phase the disease may have developed over months or years. The symptoms may be noticed as the disease transitions into the late phase.

In the late phase:
- carotid and vertebral artery dysfunction leading to ocular problems
- claudication
- diminished pulses
- vascular bruits
- aortic dysfunction
- pulmonary hypertension
- cardiomyopathy
- angina and myocardial infarction
- systemic hypertension, due to renal artery stenosis and baroreception problems

(b) Work up and treatment
- Congenital or autoimmunity factors. Perhaps *Streptococcus* or *Mycobacterium* infection, with a false-positive TB test
- Age: women aged 20–40
- Possible hypertension

Chinese Medicine

Active phase

See Figures 18.2 and 18.3.

(a) Clinical presentation. Joint and limb pain, sweating, fever, headaches, dizziness.

Tongue = red w/yellow coating

Pulse = rapid pulse

Toxic phlegm heat

Exterior pathogenic factor which quickly raises the body temperature and transforms the body fluid into stagnant phlegm which houses and replicates pathogens.

FIGURES 18.2

(b) Treatment
- Therapeutic principle:
  Clear heat, resolve phlegm, relieve pain
- Medicine:
  Si miao yong an tang
  Da qing dao tang
  Wen yang tong mai tang
  Tong mai huo xue tang

Stable phase
See Figure 18.4.

Clinical presentation. Dizziness, headache, tinnitus, lower back pain and weakness, limb numbness.
Tongue = dry coat
Pulse = rapid

See Figure 18.5.

Vessel bi syndrome

This syndrome is characterized by obstruction, narrowing or dilation of the vessels causing adverse blood flow and local pain.

FIGURE 18.3

Hyperactive liver leading to rising yang and kidney deficiency

Hyperactivity of liver yang happens as the result of prolonged emotional upset. The excessive energy (yang) consumes blood and body fluids (yin). The result is the possible stirring of endogenous wind and heart yin deficiency leading to hypertension.

FIGURE 18.4

Qi deficiency and blood stasis

Exterior pathogenic factors causing heat during the active stages have depleted qi and yin.

FIGURE 18.5

TABLE 18.4

Medicine	Action
Corticosteroids	Help control symptoms during the inflammatory process
Cytotoxic agents	Immuno-suppressant medication combined with steroids for advanced stages when relapses are more frequent
TNF inhibitors	Combines with corticosteroids and cytotoxic agents during remission stages to decrease immune responses
Antiplatelets	Inhibits platelet aggregation causing thrombosis, and renal failure
Anticoagulants	Reduces or blocks further thrombosis
Calcium channel blocker	Treats hypertension

(a) Clinical presentation. Chest suppression, breathing difficulties, dizziness, insomnia with dream-disturbed sleeping, limb pain and weakness.
    Tongue = thin white coating
    Pulse = thin to no pulse
(b) Treatment. Therapeutic principle: soothe the liver, tonify qi, break blood stasis and nourish blood.
(c) Medicine:
    Tong mai huo xue tang
    Bu yang huan wu tang
    Bu pian yu feng tang

MEDICINES

A. Western Medicine Drugs

See Table 18.4.

B. Chinese Medicine Formulas

See Table 18.5.

SUMMARY IN THE CONTEXT OF PATIENT PERSPECTIVES FOR PERSONALIZING MEDICINE

When integrating Chinese medicine and Western medicine, it is necessary to realize the side effects and complications caused by the pharmaceutical medications. In some cases the strategy for integrative medicine treatment may include using Chinese medicine during the early and pre-pulseless phases.

TABLE 18.5

Medicine	Action
Si miao yong an tang	Clears heat, detoxifies and activates blood
Wen yang tong mai tang	Activates blood, relieves pain, invigorates deficiency
Tong mai huo xue tang	Activates blood and removes stasis, regulates flow of qi, replenish blood
Bu yang huan wu tang	Invigorates qi, resolves blood stasis
Bu pian yu feng tang	Replenishes qi, soothes liver and stops wind

Western medicine can be used during the active and late phases. Chinese medicine can be used again during the stable phase, during remission. The patient narrative will help health practitioners to realize what the patient is going through. Since it has been documented about inflammation involved in this disease. Chinese medicine formulations can be provided to address aspects during remission or active phases.

FURTHER READING

[1] Feng T, Wang Z, Zhang S. Diagnosis and treatment of Takayasu arteritis. Chin J Hypertens 2008;Jun. 16(6).

[2] Ma Z, Huang X, Zhao K, Zhang S. Wenyangtongmai decoction for treatment of Takayasu arteritis in 30 Cases. Xanxi Chin Med 2013;34(2): 1000–7369(2013)02-0162-02.

[3] Yi M. The five way of Chinese medicine to Takayasu's arteritis. J N Chin Med 2009.

[4] M. Zhen, H. Xiaoli, Z. Kun, Z. Suqing (2013) Wen Yang Tong Mai decoction Tong for treatment of Takayasu arteritis in 30 cases. Shan Xi Chin Med.

[5] Xiaodong W, Huimin Y. Professor Zhang Fengshan's experience in treatment of Takayasu Arteritis. Acta Chin Med Pharmacol 2012;40(2).

[6] Xianyi L, Ping S, Feng Q, Yin W. Mechanism of vascular endothelial injury in patients with Takayasu's arteritis. J Harbin Med Univ 2010;44(5).

[7] Wei S, Meimei L, Huanping M. Clinical analysis of 47 patients with Takayasu's arteritis. J Clin Pulm Med 2013;18(5).

[8] Wenpeng J, Ren L. Combination of traditional Chinese and Western medicine treatment of brachiocephalic Takayasu's arteritis clinical observation of 18 cases. J Sichuan Tradit Chin Med 2011;29(8).

[9] Jun M, Qinghua W. Takayasu activity monitoring and decision. J Cardiovasc Pulm Dis 2010;29(3).

[10] Bujun Y, Zhanfeng H, Jianjun Z, et al. Clinical observation of arteritis renal hypertension treated with combination of traditional Chinese and Western medicine. J Tradit Chin Med 2010;51(Suppl. 1): [1] Shi Zhenyu.

[11] Li C, Wu X. The value of ultrasound diagnosis of Takayasu arteritis. Gansu Med J 2013;32(2): 1004–2725 02-0109-03.

[12] Luo S, et al. Dual-source CT angiography for the diagnosis and follow-up of Takayasu arteritis. Journal of Medical Postgraduate 2011;24(12): December; 1008–819912-1249-05.

[13] Cui Y, Liu B, Sun S. Gongrang Cui' experience in the treatment of Takayasu arteritis. Liaoning Chin Med Magzine 2012;39(4): 1000–1719 04-0628-02.

[14] Song W, Li M, Mei H. Clinical analysis of 47 Patients with multiple Takayasu arteritis. Clin Pulm Med 2013;May. 18(5).

[15] Zhang G, Du T. Differential treatment on Takayasu arteritis complicated by angina. Clin J Chin Med 2013;5(1): 1674–7860(2013)01-0088-01.

[16] Xiong Y, Hu G. Surgical treatment of type III Takayasu arteritis. Chin J Gen Surg 2010;19(06): 634–637; 1005–6947 06-0634-04.

[17] Arend WP, et al. The American College of Diagnosis, Screening and Treatment of Abdominal, Thoracoabdominal and Thoracic Aortic Aneurysms 326 Rheumatology 1990 criteria for the classification of Takayasu arterititis. Arthritis Rheum 1990;33:1129–34.

[18] Ishikawa K. Diagnostic approach and proposed criteria for the clinical diagnosis of Takayasu's arteriopathy. J Am Coll Cardiol 1988;12:964–72.

[19] Kulkarni TP, D'Cruz IA, Gandhi MJ, Dadhich DS. Reversal of renovascular hypertension caused by nonspecific aortitis after corticosteroid therapy. Br Heart J 1974;36:114–6.

[20] Cidi M, et al. Estrogens and the vascular endothelium. Ann N Y Acad Sci 2006;24 (Jan):10.1111/j.1749-6632.2002.tb04211.x.

[21] Saruhan-Direskeneli G, et al. Identification of multiple genetic susceptibility loci in Takayasu arteritis. Am J Human Genet 2013;93(2): July 298–305.

PART VI

Traditional Chinese Cardiovascular Herbs

Chapter 19

Herbs Used in Clinical Treatment

CARDIOVASCULAR

1. Heart Rate

Accelerates

- Ma huang
- She xiang
- Yangjin hua
- Cha ye

Decelerates

- Yangzhi zhu
- Dang gui
- Yu zhu
- Tusu zi
- Xucheng qing
- Qu mai
- Juli xiang

Anti-Arrhythmic

- Gan cao
- Ren shen
- Sheng di huang
- Maimen dong
- Chishao yao
- Chai hu
- Gui zhi
- Yin chen hao

2. Blood Pressure

Lowers

- Ge gen

Integrative Cardiovascular Chinese Medicine. DOI: http://dx.doi.org/10.1016/B978-0-12-420014-2.00016-8

- Luo fu mu
- Xiatian wu
- Choutou weng
- Xixian cao
- Tian ma
- Gou teng
- Shijue ming
- Huang qin
- Mu xiang
- Ye juhua
- Lianjiao
- Xiaku cao
- Hai hua
- Xuan shen
- Dan pi
- Zhi zi
- Gao ben
- Yimu cao
- Ze xie
- Che qian cao
- Che qian zi
- Dan shen
- Qu mai
- Huang qi
- Dang shen
- Sang bai pi
- Chuan xiong
- Suanzao ren
- He shou wu
- Sangji sheng
- Du zhong
- Du huo
- Wutong ye
- Yin yang huo

3. Dilates Arteries

Renal Arteries

- Huang qi
- Du zhong
- Luobu ma
- Zhi zi

Cerebral Arteries

- Ge gen
- Yin xing ye
- Qiang huo

4. Blood Circulation

Skin Circulation

- Huang qi
- Dan shen
- Rou gui
- Gui zhi
- Su ye
- Jing jie
- Niubang zi
- Sheng jiang

Reduces Capillary Permeability

- Huaimi
- Huai hua
- Lian qiao
- Huang qi
- Hong tong
- Qin jiao
- Wujia pi
- Ju pi
- Qing pi

5. Palpitations

Restless Heart and Mind

- Zhu sha
- Hu po
- Long chi
- Long gu
- Mu li
- Yuan zhi
- Fu shen
- Hehuan hua
- Yejiao teng

Insufficient Heart and Spleen

- Huang qi
- Dang shen
- Gan cao
- Dang gui
- Shoudi huang
- Suanzao ren
- Dan shen
- Heshou wu
- Da zao

Fire Due to yin Deficiency

- Shengdi huang
- Xuan shen
- Tianmen dong
- Yu zhu
- Dong qin
- Wuwei zi
- Lianzi xin
- Zhi mu

Deficiency of Heart Yang

- Gui zhi
- Long gu
- Mu li
- Fu zi
- Rou gui
- Gan jiang
- Gan cao
- Yuan zhi

Heat Attacking Heart Meridian

- Zhu ye
- Zhi zi
- Huang lian
- Shengdi huang
- Yu jin
- Dan Shen

6. Hypertension

Lowers

- Muxiang

- Che qian zi
- Luxiancao
- Hong hua
- Long kui
- Ju hua
- Shendi huang
- Mudan pi
- Cu
- Huang lian
- Xianku cao
- Xian mao
- Huang qin
- San qi
- Dan gui
- Long dan cao
- Yan hu suo
- Ren shen
- Du zhong
- Wu zhu yu
- Qiaomai
- Madouling
- Liu ye
- Lai fu zi
- Gou teng
- Shan zha
- Ge gen
- Ze xie
- Shi jue ming zi
- Niuxi
- Ci shi
- Daizhe shi
- Dan shen

7. Angina Pectoris

- Xi xin
- Han fangji
- Pu huang
- Chuan xiong
- San qi
- Yanhu suo
- Wuling zhi
- Dan shen
- Gui zhi

8. Cardiogenic Shock Recovery

- Zhi shi
- Qing pi
- Fu zi
- Gan jiang
- Ren shen

9. Thromboangiitis Obliterans

- Hung hua
- Fu zi
- Mantuoluo hua
- Zao xiu
- Dan shen
- Zi cao
- Maodong qing
- Dang gui
- Song xiang
- Tao ren
- Jixue teng
- Chuan xiong
- Chuanshan jia

10. Rhythm Disorders

Arrhythmia

- Yu jin
- Fu zi
- Dan shen
- Yanhu suo
- Ku shen
- Shangji sheng
- Shandou gen
- Ge gen
- Long yan rou
- Gan cao
- Da zao
- Ku shen

Sinus Tachycardia

- Cangzhu
- Qing pi
- Fang ji

Sinus Bradycardia

- Fu zi
- Xi xin
- Huang jin
- Gui zhi
- Bugu zhi

Extra Systole

- Hong hua
- Yin chen hao
- Huang qi
- Gan cao
- Yanhu suo
- Ku shen
- Bai fan
- Xiaonie

AV Block

- Fu zi
- Lu rong

11. Chest Pain Relief

- Gualou pi
- Ban xia
- Gui zhi
- Zhi shi
- Bi ba
- Tan xiang
- Gaoliang jiang
- Jiang xian
- Chen xiang
- Dan shen
- Hong hua
- Maodong qing
- Chuan xiong
- Tao ren

12. Heart Yang Deficiency

- Gui zhi
- Rou gui
- Fu zi

- Xie bai
- Gan jiang
- Da zao

13. Heart qi Deficiency

- Ren shen
- Dang shen
- Huang qi
- Fu ling
- Wu wei zi
- Gan cao
- Taizi shen
- Da zao

14. Heart Blood

Heart Blood Deficiency

- Dang gui
- Baishao yao
- Jixue teng
- Longyang rou
- Zihe che
- Shoudi huang
- He shou wu

Heart Blood Stagnation

- Dan shen
- Tao ren
- Hong hua
- Chuan xiong
- Chishao yao
- Yu jin
- Maomen don
- Bai guo
- Tanxiang
- Chen xiang
- Ding xiang
- Ru xiang
- She xiang
- Suhe xiang

15. Heart yin

- Shengdi huang
- Maimen dong
- Yu zhu
- E jiao
- Bai he
- Wu wei zi
- Suanzao ren
- Xiyang shen

16. Heart shen

Tranquilizing

- Suanzao ren
- Heshou wu
- Yuan zhi
- Hehuan pi
- Hehuan hua
- Maimen dong
- Hu po
- Long gu
- Long chi
- Cishi
- Zhen zhu mu
- Mu li

Excessive Heart Fire

- Zh sha
- Niuhuang
- Huang lina
- Zhi zi
- Lianzi xin
- Zhu ye
- Lian qiao
- Shengdi huang
- Mu dan pi

Phlegm Blocking the Heart Orifices

- Yuanzhi
- Yujin
- Xi xin
- Zhuy zhao
- Zhu li

RESPIRATORY

1. Anti-Anaphylaxis

- Wu mei
- Di long
- Huang qi
- Han fangji
- Zhen zhu
- Mudan pi
- Shi wei
- Gan cao
- Qin jiao
- Chai hu
- Cang zhu
- Shi wei
- Zhen zhu
- Yejiao teng
- Baijili
- Ren shen
- Fu ping
- Jing jie
- Ren dong teng
- Qiang hup
- Xiheliu
- Zhi zi
- Can sha
- Baixian pi

2. Dyspnea and Orthopnea

Excess Syndrome and Reversal of qi Flow

- Ma huang
- Su zi
- Tingli zi
- Xing ren
- Hou po
- Chen xiang
- Yangjin hua

Excess Syndrome and Fluid Retention

- Tinglizi
- Lai fuzi
- Xuanfu hua

- Ban xia
- Gan jiang
- Xi xin
- Sang bai pi
- Shi wei

Deficiency of Lung qi

- Wu wei zi
- He zi
- Wu mei
- Bai guo

Deficiency Due to Kidney Failure to Receive qi

- Fu zi
- Rougui
- Shou di huang
- Huang jing
- Wu wei zi
- Gou qi zi
- Bu gu zhi
- Chen xiang
- Ge jie
- Ci shi
- Mian hua gen

GASTROINTESTINAL/HEPATOBILIARY

1. Reduces Transaminase

- Shuifei ji
- Wu wei zi
- Ling zhi
- Dang yao
- Long dan cao
- Dan shen
- Chai hu
- Chuipen cao
- Lian qiao
- Da qing ye
- Ye ju hua
- Xi xian cao

2. Relaxes Gastrointestinal Tissue

- Yangjin hua
- Yingsu qiao
- Rou gui
- Chen xiang
- Wu zhu yu
- Huo xiang
- Wu yao
- Zhe bei mu
- Huang qin
- Qing mu xiang
- Xiao hui xiang
- Hu lu ba
- Ju pi
- Gao liang jiang
- Chi shao yao
- Gan cao
- Shi chang pu
- Baitou weng
- Ma huang
- Niu xi
- Lao guan cao
- Xu chang qing
- Huang jing
- Dongchong xia cao
- Tian xian zi
- Chi shao yao

3. Stimulates Gastrointestinal Tissue

- Da huang
- Zhi shi
- Mang xiao
- Bin lang
- Lai fu zi
- Mu xiang
- Sha ren
- Caodou kou
- Sheng jiang
- Su ye
- He shou wu
- Hu sui
- Sheng ma

- Da fu pi
- Shi hu
- Hou po
- Wu yao

NEPHROLOGY

1. Nephrotic

Diuretic (Edema)

- Zhu ling
- Ze xie
- Cang zhu
- Bai zhu
- Che qian cao
- Che qian zi
- Dan zhu ye
- Hu po
- Qu mai
- Ban bian lian
- Long kui
- Hai jin sha
- Bi xie
- Shi wei
- Lu gen
- Don gua pi
- Dan zhu ye
- Hu po
- Bian xu
- Ba bian lian
- Long kui
- Hai jin sha
- Shi wei
- Yu mi xu
- Lu gen
- Dong gua ppi
- Xia ku cao
- Ma huang
- Xiang ru
- Ya zhi cao
- Ku shen
- Huang qin
- Yi mu cao
- Da fu pi

- Fang ji
- Tingli zi
- Jia hu tao
- Sha zhu yu
- Gan sui
- Huang qi
- Sang ji sheng
- Da ji
- Yuan hua
- Qian niu zi
- Shang lu
- Hu lu
- Liaoge wan gen
- Shanbian dou
- Xuan fu hua
- Shi xie
- Jingu cao
- Yizhihuang hua
- Banbian lian

Reduces Hyperproteinuria

- Huang qi
- Ren shen
- Bai zhu
- Fu ling
- Bie jia jiao
- Shan yao
- Dang gui
- Gou qi zi
- Jinying zi
- Sangpiao xiao
- Lian xu
- Niu xi
- Du zhong
- Shengdi huang
- Xuan shen
- Mai men dong
- Tusi zi
- Tufu ling
- Su ye
- Yimu cao

Reduces Hyperlipidemia

- Jue mingzi
- Yin chen hao
- Che qian cao
- Yinxing ye
- Ling zhi
- Heshou wu
- Du zhong
- Wu tong ye
- Sangji seng
- Ju hua
- Yu zhu
- Huang jing
- Qianshi
- Jin yin hua
- Gouqi zi
- Huang jin
- Qian shi
- Huang qi
- Gang gui
- San qi
- Hua jiao
- Yu jin
- Kun bu
- Jiang huang
- Shi shu ye
- Chashu gen
- Jin yin hua

Glomerulonephritis

- Cheqian cao
- Tingli zi
- Deng xin
- Ma huang
- Yu jin
- Banbian lian
- Haijin sha
- Lian qiao
- Mabian cao
- Huang qin
- Huang qi

- Baimao gen
- Yimu cao
- Laliao cao
- Hu jiao
- Jinyin hua
- Renru zhi
- Da suan
- Shi wei
- Fu ping

2. Nephritic

These herbs are either not addressed in this volume of integrative medicine or are found in the relevant herb categories where they concern certain symptoms of hypertension, oliguria, azotemia and hematuria.

Chronic Nephritis

- Yi mu cao
- Shuiyang mei
- Chuan xiong
- Shui zhi

Chronic Renal Failure

- Da huang
- Chiqian zi
- Hua shi
- Huai jiao
- Machixian
- Bian xu
- Qu mai
- Hai jin sha
- Shi wei
- Bi xie

IMMUNE SYSTEM

1. Promotes Immune System

These herbs can be used for supplementing cardiovascular formulas during exterior syndromes.

- Ren shen
- Huang qi
- Yin yang huo

- Sang zhi
- Wujia pi

2. Endoreticular System Stimulation

These herbs can be used for supplementing cardiovascular formulas during exterior syndromes.

- Jin yin hua
- Huang lian
- Huang bai
- Da qing ye
- Banlan gen
- Bai hua she she cao
- Chuan xin lian
- Shan dou gen
- Yuxing cao
- Yizhihuang hua
- Huang qi

3. Anti-Inflammatory

These herbs can be used for supplementing cardiovascular formulas during exterior syndromes to prevent or treat complications of infection.

- Shi gao
- Huang qin
- Long dan cao
- Ku shen
- Jin yin hua
- Da qing ye
- Bai hua she she cao
- Ban zhi lian
- Chuanxin lian
- Sha shen
- Shi hu
- Yu zhu

4. Anti-Pyretic

These herbs can be used for supplementing cardiovascular formulas during exterior syndromes.

(a) Regulates Heat

- chai hu
- huang qin

- zhi mu
- shi gao
- qing hao
- yin chen hao
- mudan pi
- huang lian
- xi xin
- zhi zi
- ju hua
- feng fang
- yin chai hu
- zhi cao
- weiling xian
- dan zhu ye
- qian hu

(b) Induces sweating

- ma huang
- gui zi
- xiang ru
- su ye
- jing jie
- fang feng
- qing jiao
- niubang zi
- chai hu
- sheng ma
- qin jiao
- ge gen
- fu ping
- baixiang pi

ANTIMICROBIALS

1. Antibacterial

These herbs can be used for supplementing cardiovascular formulas during exterior syndromes.

Broad-Spectrum Antibacterials

- Huanglian
- Huangqin
- Huangbai

- Daqing ye
- Chuanxin lian
- Mudan pi

Anti-Staphylococcus

- Jie geng
- Bai tou weng
- Xian he cao
- Qian cao
- Da huang
- Qu man
- Bian xu
- Niubang zi
- Hou po
- Canger zi
- Xuan shen

Anti-Streptococcus

- Yuxing cao
- Canger zi
- Ai ye

Rheumatic Fever and Endocarditis

These herbs can be used for supplementing antibiotic formulas.

- Qing hao
- Huang qin
- Fang ji
- Rendong fang
- Qin jiao
- Huang bai
- Quan xie
- Wushao she

Anti-Haemophilus Influenzae

- Banlan gen
- Jie geneg
- Bai zhi
- Gua lou
- Wu wei zi

Anti-Mycobacterium Tuberculosis

- Huanglian
- Huang bai

- Ku shen
- Xiaku cao
- Jin yin hua
- Chai hu
- Yin chen hao
- Dan shen
- Yu zhu

Anti-Bordetella Pertussis

- Huang qin
- Chuanxin lian
- Hou po
- Xiao ji

Anti-Bacillus *Enteritis*

- Diyin cao
- Xianhe cao
- Chi shao yao
- Jing jie
- Qing hao

Anti-Bacillus *Dysentery*

- Ku shen
- Bai tou weng
- Yu xing cao
- Mu xiang
- Shi liu pi
- Dang gui

2. Antiviral

These herbs can be used for supplementing cardiovascular formulas during exterior syndromes.

Anti-Influenza

- Da qing ye
- Ban lan gen
- Huang qin
- Huang lian
- Huang bai
- Niu bang zi
- Mu dan pi
- Yin chen hao
- Huang jin

Anti-Coxsackievirus

- Huang lian
- Da qing ye
- San ji sheng
- Pu gong yin
- Chuanxin lian
- Ye ching shao

3. Anti-Protozoan

These herbs can be used for supplementing cardiovascular formulas during exterior syndromes.

Anti-Plasmodium

- Chai hu
- Yadan zi
- Huang qin
- Huang lian
- Huang bai
- Cang zhu
- Longdan cao
- Sheng ma

Anti-Amoeba

- Baitou weng
- Ku shen
- Huang lian
- Huang bai
- Ya dan zi
- Weixin lian

4. Anti-Spirochaete

These herbs can be used for supplementing cardiovascular formulas during exterior syndromes.

- Daqing ye
- Ban lan gen
- Chuanxin lian
- Huang lian
- Huang qin
- Huang bai
- Lian qiao
- Zhi zi
- Tu fu ling

5. Anti-*Candida Albicans*

These herbs can be used for supplementing cardiovascular formulas during exterior syndromes.

- huang bai
- huang qin
- huang lian
- chuanliang zi
- shandou gen
- tujin pi

6. Anti-Parasitic

These herbs can be used for supplementing cardiovascular formulas during exterior syndromes.

Anti-Ascaris

- kulian pi
- chuanlian pi
- wuzhu yu
- hou po
- shijun zi
- qianniu zi

Anti-Schizoneptus

- xiaohuixiang
- danshen
- qu mai
- zhi zi

Anti-Filiaris

- weixin lian
- qing hao
- sang ye
- beiwujia pi

ENDOCRINE

1. Pituitary-Adreno-Cortical Stimulation

- Fu zi
- Wu tou
- Ren shen

- Renshen ye
- Qin jiao
- Han fang ji
- Jinguo lan
- Mian hua gen
- Wuwei zi
- Feng du
- Chuanshan long
- Qingfeng teng
- Feng du
- Chuanshan long
- Shisuan
- Shuiniu jiao
- Gan cao
- Huang qi
- He shou wu
- Yu zhu
- Ciwu jia

2. Increases Iodine

- Kunbu
- Hai zao
- Hai dai
- Zi cai

3. Promotion and Increase of Metabolism

- Ren shen
- Ma huang
- Cha ye
- Niubang zi

4. Diabetes

Reduces Concentration of Blood Sugar

- Ren shen
- Huang qi
- Fu ling
- Bai zhu
- Cang zhu
- Shan yao
- Huang jing
- Sheng di huang

- Shou di huang
- Xuan shen
- Maimen dong
- Zhi mu
- Tianhua fen
- Yu zhu
- Gouqi zi
- Heshou wu
- Wuwei zi
- Yinyang huo
- Fengru
- Zexie
- Yumixu
- Digupi
- Hu zhang
- Xianhe cao
- Wujia pi
- Canger zi
- Wubei zi
- Tianmen dong

Increases Concentration of Blood Sugar

- Dang shen
- Si ye shen
- Shi hu
- Huang qin
- Qin jiao
- Zhu ye
- Sheng jiang
- Huai hua

BLOOD

1. Increases Erythrocytes and Hemaglobin

- Lu rong
- Lu jiao
- Zihe cao
- Ejiao
- Jixue teng
- Ren shen
- Huang qi
- Dang shen
- Heshou wu

- Siye shen
- Dang gui
- Shou di huang
- Gouqi zi
- Bai zhu
- Yejiao teng
- Long yan rou
- Bugu zhi
- Suo yang
- Baiji tian
- Ju pi
- Dan shen

2. Increases White Blood Cells

- Ren shen
- Jing gu cao
- Jixue teng
- Dan shen
- Chuanshan jia
- Hu zhang
- Shi wei
- Ru xiang
- Mo yao

3. Reduces White Blood Cells

- Xi jiao
- Dan shen
- Siye shen
- Tubie chong
- Hu zhang

4. Increases Reticulated Corpuscles

- Jinyin hua
- Huang lian
- Huang bai
- Daqing ye
- Banlan gen
- Baihua she she cao
- Chuanxin lian
- Shandou gen
- Yuxing cao
- Yizhihuang hua
- Huang qi

5. Platelet Elevation

- Dang gui
- Bai shao yao
- Sheng di huang
- Shou di huang
- Shanzhu yu
- Longyan rou
- Da zao
- Da huang
- Niu xi
- Yang ti
- San qi
- Bai ji
- Ou jie
- Xian he cao
- Rou cong rong
- Gou ji shuiniu jiao
- Huang bai
- Lian qiao

6. Platelet Reduction

- Huang qi
- Taizi shen
- Bai zhu
- Dang gui
- E jiao
- Chuan shanjia
- Dan shen
- Jixue teng
- Sheng di huang
- Shou di huang
- Dongchongxia cao
- Gou qi zi
- Wu wei zi
- Shan zhu yu
- Bu gu zhi
- Nuzhen zi
- Shi wei
- Ling zhi
- Shi hu

- Yi zhi ren
- She chuang zi
- Mo gu

7. Stops Bleeding

- San qi
- Xue jie
- Mo han lian
- Xian he cao
- Bai ji
- Mu dan pi
- Zhi zi
- Jinggu cao
- Cebai ye
- Baimao gen
- E jiao
- Yuxing cao
- Jiubi ying
- Niu xi
- Yang ti
- Qian cao
- Dijin cao
- Di yu
- Huai hua
- Huai jiao
- Pu huang
- Ai ye
- Guan zhong
- Wu bei zi
- Gou ji
- Wu zei gu
- Hua rui shi
- Da ji
- Xiao ji
- Ji cai
- Du zhong
- Bugu zhi
- Roucong rong
- Zi zhi
- Ou jie
- Pao jinag

- Hiang qin
- Jing jie
- Huang lian
- Wu mei
- E jiao

8. Anticoagulant

- Shui zhi
- Hai zao

9. Anemia

Iron-Deficiency Anemia

- Ren shen
- Dan gui
- Sang shen
- Long yan rou
- Bai shao yao
- Shou di huang

Aplastic anemia

- Huang qi
- Lu rong
- Ren shen
- Dang gui
- Shou di huang

Polycythemia Vera

- Yimu cao
- Shui zhi
- Xiong huang
- Dan shen

Reduces Thrombocytopenia Purpura

(This is a condition often caused by infection, pharmaceutical medication for autoimmune diseases, blood clots and hormone replacement.)

- Luxian cao
- Muzei
- Shang lu
- Lian qiao
- Chai hu

- Zi cao
- Suo yang
- Gan cao
- Huang qi

PALLIATIVE

1. Insomnia

- Yan hu suo
- Ku shen
- He huan pi
- E jiao
- Suan zao ren
- Long chi
- Huang lian
- Xiao mai
- He shou wu
- Yejiao teng
- Hu po
- Long gu

2. Headache

- Niu xi
- Tian ma
- Dang gui
- Chuan xiong
- Xi xin
- Chan tui
- Bai guo
- Qiao mai
- Ge gen
- Ding xiang
- Shi yan
- Bai zhi
- Fane feng
- Gou teng
- Tu fu ling
- Sang ye
- Ju hua
- Jing jie

3. Sedatives

- Suan zao ren
- Yan hu suo
- Ling zhi
- Dang gui
- Chuan xiong
- Fu shen
- Tian ma
- Gou teng
- Manjing zi
- Gao ben
- Zhi zi
- Lianzi xin
- Chan tui
- Di long
- Tian ma xing
- Zhen zhu
- Bai ji li
- Su mu
- Shu sha
- Niu huang
- Long gu
- Yejiao teng
- Bia zi ren
- He huan pi
- Chai hu
- Huang qin
- Zhi mu
- Du huo
- Xiang fu
- Xi xian cao
- Chou wu tong
- Xixian cao
- Qin pi
- Qin jiao
- Du zhong
- Baiji tian
- Gou qi zi
- Fu xiao mai
- Long yan rou
- Tian zhu huang
- Huang yang jiao

- Tai zishen
- Luo bu ma
- Shi chang pu
- Xu chang qing
- Qi ye yi zhi hua
- Bai hua she she cao

PART VII

Integrating TCM and Western Medicine

Chapter 20

Clinical Management of Cardiovascular Formula and Drug Therapy

INTRODUCTION

In Chinese medicine, prescribing medicine is generally the result of diagnosis using syndrome differentiation and later setting a modifiable treatment principle. The 'ben', or chief disease, and the 'biao', or secondary symptoms, are considered, and they help to make decisions about the direction of the course of treatment. Patent medicines are provided, or formulas are arranged with the careful selection of herbs from various or particular categories of the Chinese Pharmacopeia. Any changes observed or reported in the patient's condition are noted in subsequent visits. Modifications in syndrome differentiation, and the prescribing of medicine, continue on the basis of tongue and pulse diagnosis.

Integrative Chinese medicine for the management of cardiovascular diseases is similar to a subspecialty. It includes aspects of Chinese medicine and Western medicine, basic medical sciences and theories, perspectives in pathology, examination, diagnosis, prescribing, prognosis and management. Diseases and syndromes are observed as both common among patients and unique to a given patient. As a disease unfolds, symptoms develop and become established. Adjusting to, and coping with, symptoms can often become part of a patient's way of living. Patterns emerge with changes in the season, or due to emotional circumstances, environmental exposure and according to the patient's constitution. The methods involved in the art of prescribing integrative medicinal prescriptions are of particular interest. Not only do they require understanding of the herbs, formulas and drugs used, they also require a knowledge of the general and specific characteristics of cardiovascular diseases. For health practitioners initiating and establishing integrative Chinese medicine for the treatment of patients with cardiovascular diseases, there are three methods which can guide further development in clinical management; these are discussed below.

The Basic Method for Retaining or Referring

Health practitioners must maintain the belief that the needs of the patient are more important than personal bias towards or against certain medical systems.

Integrative Cardiovascular Chinese Medicine. DOI: http://dx.doi.org/10.1016/B978-0-12-420014-2.00017-X

For example, the control of blood pressure can be complicated. The use of herbal formulas may provide more favorable results where pharmaceutical drugs may not, and some patients require pharmaceutical drug therapy to control symptoms. Herbal formulas require a sense of responsibility for the patient to prepare, ingest and observe themselves while under the effects of these formulas. For the most part, a properly prescribed formula should take effect within the day it was dispensed and ingested. A patient's condition not responding to a herbal formula within 3–5 days should be reviewed immediately. The patient may be non-compliant, the formula may not be properly prescribed, or the condition may be more serious than originally thought and requires referral to outside assistance.

The Basic Method for Indicating or Contraindicating

While making a decision whether to prescribe exclusively or integrate herbs and drugs, it is necessary to compare and contrast their actions and indications. Pharmaceutical drug therapy may sometimes be necessary in some patients. For them, certain raw herbal decoctions may complicate the course of treatment. Conversely, many raw and patent herbal formulas can favorably guide the patient's condition to a more robust and developed recovery, where pharmaceutical drugs may facilitate other adverse symptoms and complicate recovery. Combining herbal formulations and pharmaceutical drugs often takes thoughtful analysis of the patient's overall condition, along with an understanding of pharmacokinetics and predicting the pharmacodynamics. In addition, a basic idea of how to develop a method of integrating is to observe the patient on Chinese medicine to cover certain disease conditions which may have to be overlooked when certain pharmaceutical drugs are prescribed. For example, if a patient with heart disease suddenly develops a cold or the flu, it is imperative to relieve the condition as soon as possible to avoid complications. Chinese medicine and Western medicine may have contrasting ideas about how to proceed. The Western medicine practitioner may simply prescribe or suggest medicines along with the cocktail of drugs already being taken by the patient. Whereas in Chinese medicine, some practitioners may believe 'gan mao' is a situation when tonifying herbs or formulas are suddenly contraindicated, and attention must be paid to relieving the exterior, the main treatment being resumed later. Some practitioners may guide out the exterior condition with the addition of certain herbs to existing cardiovascular formulas without discontinuing treatment. Others may prescribe herbal formulas or single herbs which can independently address a new condition 'quietly' while the pharmaceutical drugs continue to address the chief cardiovascular disease.

At times, combining herbal formulas and pharmaceutical drugs may cause more harm than good. In this case, the actions of both varieties of medicines would be similar and can increase the adverse effects. An example is a treatment involving blood thinners, such as warfarin and heparin. The use of nutritional supplements which contain co-enzyme Q_{10}, vitamin K and vitamin C has been associated with a decrease in prothrombin ratio (PR) and the international normalized ratio (INR), which means that they may cause hemorrhaging with the use of warfarin. Blood-moving formulas with herbs – such as dang shen, dang gui, ren shen, gan jiang, bai guo, hong hua, tao ren, bai guo ye and gan cao – are other examples. Basic situations in which combining herbal formulas with pharmaceutical drugs can be beneficial and include treating pathogenic factors, protecting organs during pharmaceutical drug courses, and medicinal meals to help a patient to recover after a serious illness.

The Basic Method for Prescribing Exclusively or Integrating

Health practitioners using exclusive or integrative treatments should base decisions on clear judgment from observation and physical examination. One point to note is the necessity to consider periods of cold and flu, as well as other infections, with extreme seriousness. Common factors in cardiovascular diseases where any infection is likely to create marked complications in these patients include anemia, increase in platelet count, inflammation, edema, and an excess of cholesterol and lipoprotein. Examples of organ system dysfunction which react to various microbial infections, and which are involved directly with cardiovascular function, include lung disorders, kidney and liver weaknesses or diseases, and gastrointestinal upset or diseases.

Cardiovascular disease prevention stages often involves lifestyle-modification with absence-of-symptoms, and a development period in which factors should be controlled to reduce or eliminate symptoms. During the prevention stage, Chinese medicine formulas have a leading appeal as patients may benefit from herbal formulas alone – or herbal formulas combined temporarily with pharmaceutical drugs. During critical disease stages clinical presentations range from the less benign and controllable, through periods of health debilitation, to the terminal stage. Western medicine may have the leading appeal, but Chinese herbal medicine may still be of benefit for biao secondary purposes. For example, some Chinese medicine formulas could be used to help protect the stomach, liver or kidneys during a treatment course with pharmaceutical drugs which are known to harm these organs. In addition, when a cardiovascular episode requires surgery, or a serious infection such as a complicated strain of influenza, pneumonia or other resistant microbial infection occurs during treatment, the patient can benefit from Chinese medicine. Here, certain herbal formulas, such as Wuji bai feng, for example, may not be indicated to directly target the pathologic factors, but after the critical period and during recovery, this medicine can function more like nourishment to help with re-building of qi and blood to improve the prognosis.

CLINICAL MANAGEMENT ACCORDING TO CHAPTERS

Chapter 9: Lipoprotein Disorders

1. *Pharmaceutical Drugs*

(a) Ezetimibe: lowers cholesterol absorption in the small intestine.
(b) Statins: inhibit HMG-CoA reductase enzyme for cholesterol reduction.
(c) Bile acid sequestrants: prevent fat absorption in the intestines.
(d) Fibrates: combine with statins to treat high cholesterol.
(e) Niacin: helps break down fat tissue.

2. *Chinese Medicine Formulas*

(a) Fleece flower formula. Reduces liver hyperactivity, relieves body distention, nourishes kidney yin and blood. Dosage: 8 oz/227 g or 30 tea pills 3 × daily as needed.
(b) Weicao tang. Reduces blood lipids and increases lipid metabolism. Dosage: 8 oz/227 g or 30 tea pills 3 × daily as needed.
(c) Gentian Formula. Purges liver fire to reduce adverse effects on the heart, pathogenic dampness transforming into heat. Dosage: 8 oz/227 g or 30 tea pills 3 × daily as needed.
(d) Eliminate Blood Lipid Formula. Reduces blood lipids for weight loss. Dosage: 8 oz/227 g or 30 tea pills 3 × daily as needed.
(e) Shen qi wan. Helps to tonify and strengthen the kidneys. Dosage: 8 oz/227 g or 30 tea pills 3 × daily as needed.
(f) Reduce Blood Lipid Formula. Reduces LDL levels. Dosage: 8 oz/227 g or 30 tea pills 3 × daily as needed.
(g) Ginko leaf formula. Relieves symptoms of lipoproteinemia, protects against free radicals, moves blood, and breaks up phlegm in the vessels. Dosage: 8 oz/227 g or 30 tea pills 3 × daily as needed.

Integrating Chinese Herbal Formulas and Western Medicine Drugs During Treatment

(a) Basic methods for indicating or contraindicating:
- Bile acid sequestrants prevent fat absorption in the intestine and assist in the rise of HDL levels. However, they can also block the absorption of vitamins A, D, E and K, vitamins C and E, and beta carotene. This interferes with the use of herbal formulas and nutritional supplements.
- Niacinamide is the constituent in niacin which helps in the treatment of low HDL levels. Certain problems associated with its improper use include flushing, increase in histamine, arrhythmias, blood sugar problems, low blood pressure, stomach ulcer, and liver, gall bladder and kidney disease complications.

- Fibrates increase the effect of digoxin and medications for endocrine disorders and should be taken an hour before, or at least 4 hours after, these types of medication.

(b) Basic methods for prescribing exclusively or integrating:

- Chinese medicine formulas for the treatment of lipoprotein disorders may benefit patients intolerant of pharmaceutical drugs. It is not recommended that patients blend herbal formulas with drugs owing to certain constituents, including the mineral and vitamin content of the herbs.
- Chinese medicine formulas are as effective as pharmaceutical drugs for lowering blood lipids. In some cases where a patient has been taking drugs to lower lipids, and is indicated for Chinese medicine, the patient can be weaned off the drugs and started on Chinese medicine within a few days.
- Shen qi wan, Gentian Formula, Gentian Formula and Fleece flower formula can help protect viscera with niacin use. Aspirin may reduce the flushing caused by niacin.

(c) Additional herbal modifications (see Chapter 19):

With these formulas, herbs from the hyperlipidemia category, antimicrobial category, nephrology category and endocrine category may be necessary.

Chapter 10: Hypertension

1. *Pharmaceutical Drugs*

(a) Thiazide diuretics: reduce edema.
(b) Beta blockers: heart beats slower and with less force.
(c) ACE inhibitors: relax blood vessels.
(d) ARBs: relax blood vessels.
(e) Calcium channel blockers: relax blood vessels.
(f) Renin inhibitors: slow the production of renin.
(g) Alpha blockers: reduce nerve impulses, slow the heartbeat.
(h) Central-acting agents: prevent heart rate increase and blood vessel narrowing.
(i) Vasodilators: prevent constriction of arteries.

2. *Chinese Medicine Formulas*

(a) Gambirplant Branch Formula. Reduces hypertension, headache and dizziness. Dosage: 8 oz/227 g or 30 tea pills 3 × daily as needed.
(b) Yang Hyperactivity Check with 7 Drugs Formula. Reduces hypertension, headache, tinnitus, and is mildly diuretic. Dosage: 8 oz/227 g or 30 tea pills 3 × daily as needed.

(c) Liver subduing and Wind Stopping Formula. Reduces hypertension, thyroid problems, hyperaldosteronism, and tranquilizes the mind. Dosage: 8 oz/227 g or 30 tea pills 3 × daily as needed.
(d) Blood Pressure-Reducing Decoction. Reduces blood pressure, dreaminess, insomnia, brain fog, chest oppression. Dosage: 8 oz/227 g or 30 tea pills 3 × daily as needed.
(e) Qi-Replenishing and Yin-Nourishing Formula. Stabilizes blood pressure while recovering from dryness, exhaustion and headache. Dosage: 8 oz/227 g or 30 tea pills 3 × daily as needed.
(f) Gastrodia and Uncaria Formula. Reduces blood pressure in serious cases, and coordinates with pharmaceutical drugs. Dosage: 8 oz/227 g or 30 tea pills 3 × daily as needed.
(g) Gu Chong Formula. Reduces blood pressure, especially during spikes between pharmaceutic medications. Dosage: 8 oz/227 g or 30 tea pills 3 × daily as needed.
(h) Cardiotonic Formula. Reduces blood pressure in serious cases, and coordinates with pharmaceutical drugs. Dosage: 8 oz/227 g or 30 tea pills 3 × daily as needed.

Integrating Chinese Herbal Formulas and Western Medicine Drugs During Treatment

(a) Basic methods for indicating or contraindicating:
- Chinese medicine formulas can benefit patients alone or integrated with pharmaceutical drugs for hypertension. This is best indicated for patients with mild to moderate hypertension. In patients with mild hypertension, a herbal formula alone can provide a feeling of relief and wellbeing in the chest within a few hours after taking. The general course of this initial treatment is three times daily for 14 days. This helps the patient begin to modify their lifestyle and establish justification for continuing, because they might feel the difference between, before and during the treatment.
- Patients in the moderate hypertension range, where the blood pressure seems to fluctuate within the range 140–150/90–100, may benefit more by combining Chinese and Western medicine. When the blood pressure is elevated, specific symptoms may include anxious feelings in the chest, tension in the shoulders, heightened irritability and one of the following additional symptoms: insomnia in the form of racing thoughts at night; exhaustion during the day and inability to sleep at night; and exhaustion at various times during the day and night with more wakeful energy during the middle of the night.
- Patients in the moderate to high ranges may need pharmaceutical drugs to maintain a favorable blood pressure range. Many reasons are related to the kidney, endocrine and nervous system causes. If they

can benefit from herbal formulas, it may help with relaxing during the evening to assist with sleeping at night.

- Gastrodia and Uncaria formula, Gu Chong formula and/or Cardiotonic formula are Chinese medicines in this category of use which may be prescribed alone. When prescribed along with pharmaceutical drug therapy, monitoring is necessary. General dosage is 4 hours between pharmaceutical drug doses to relieve symptoms.

It is necessary to realize that certain disease conditions may need pharmaceutical drug therapy. It is important to conduct a thorough examination to determine whether a patient is indicated for integrating herbal formulas, or modifying with other herbs which will decrease problems with interaction.

- Thiazides may cause dizziness and lightheadedness, blurred vision, loss of appetite, itching, stomach upset, headache and weakness, muscle cramps, pain, nausea, vomiting.
- Beta blockers may cause depression, vision problems, nightmares and hallucinations, diarrhea, and gastrointestinal problems such as cramping, nausea and vomiting. Other problems can include problems with interaction with other drugs for hypertension, blood glucose problems in diabetics, fatigue, and heart failure in patients already having disease symptoms.
- ACE inhibitors and ARBs have few interactions with other drugs. However, aspirin may reduce the effects. Because ACE and ARBs increase serum potassium levels, patients who use supplements, herbs and other substances which contain potassium may develop cardiac conduction problems.
- Some patients should not use pharmaceutical drugs owing to problems with the renal artery and other issues of kidney failure, decrease in white blood cells and edema.
- Calcium channel blockers have side effects in some patients – e.g. hepatic and gastrointestinal problems, edema, or low blood pressure with symptoms of dizziness and fatigue.

(b) Basic methods for prescribing exclusively or integrating:

- The method of prescribing for hypertension relief is similar to treating the disharmony between the heart yang and kidney yin. A herbal formula containing herbs which tonify qi and/or yang can be regularly alternated with a herbal formula with herbs that tranquilize.
- Temporarily integrating herbal formulas with pharmaceutical drugs may require a similar rhythm in some patients. This rhythm may last for 5 to 7 days to either wean or transition therapies. For example, a patient may have been on pharmaceutical drug treatment and may be ready to move towards the prevention stages in Chinese medicine; beta blockers, for example, should not be withdrawn suddenly because of a return of symptoms or sudden death.

- In addition, abruptly stopping ACE inhibitors and ARBs may cause instability in blood pressure. For patients who are indicated for Chinese medicine as an alternative to pharmaceutical drugs, Gambirplant Branch Formula, Yang Hyperactivity Check with 7 Drugs Formula, Liver-subduing and Wind-stopping Formula, Blood Pressure-Reducing Decoction and Qi-Replenishing and Yin-Nourishing Formula can be used.

  Effects such as a feeling of calmness and peace of mind, as well as reduction in blood pressure, can be expected within 30 minutes to 2 hours after dosage. These medicines are best used in conjunction with other lifestyle modifications.

(c) Additional herbal modifications (see Chapter 19).

With these formulas, herbs from the cardiovascular category, palliative category, antimicrobial category, blood category, nephrology category and endocrine category may be necessary.

Chapter 11: Arrhythmia

1. *Pharmaceutical Drugs*

(a) For patients without accessory pathway:
 Esmolol: Class I, evidence C
 Metroprolol: Class I, evidence C
 Propanolol: Class I, evidence C
 Diliatazem: Class I, evidence B
 Verapamil: Class I, evidence B

For patients with accessory pathways:
 Amiodarone: Class IIa, evidence C
 Digoxin: Class I, evidence B

(b) For patients with heart failure without accessory pathways:
 Digoxin: Class I, evidence B
 Amiodarone: Class I, evidence C

2. *Chinese Medicine Formulas*

(a) An shen ning xin Formula. Soothes the spirit. Dosage: 8 oz/227 g or 30 tea pills 3 × daily as needed.

(b) Fu mai decoction. Qi and blood deficiency. Dosage: 8 oz/227 g or 30 tea pills 3 × daily as needed.

(c) Shen Fu decoction. Helps the patient to recover after illness, warms yang. Dosage: 8 oz/227 g or 30 tea pills 3 × daily as needed.

(d) Si Ni decoction. Helps the patient to recover after illness, warms yang. Dosage: 8 oz/227 g or 30 tea pills 3 × daily as needed.

(e) Wendan decoction. Regulates qi and transforms phlegm. Dosage: 8 oz/227 g or 30 tea pills 3 × daily as needed.
(f) Linggui Shugan decoction. Phlegm retention. Dosage: 8 oz/227 g or 30 tea pills 3 × daily as needed.
(g) Xuefu Zhuyu decoction. Blood stagnation and stasis. Dosage: 8 oz/227 g or 30 tea pills 3 × daily as needed.
(h) Taohong Siwu decoction. Invigorates blood. Dosage: 8 oz/227 g or 30 tea pills 3 × daily as needed.
(i) Zhi gancao decoction. Helps the patient to recover after illness, regulates heart. Dosage: 8 oz/227 g or 30 tea pills 3 × daily as needed.
(j) Guipi decoction. Nourishes blood. Dosage: 8 oz/227 g or 30 tea pills 3 × daily as needed.

Integrating Chinese Herbal Formulas and Western Medicine Drugs During Treatment

(a) Basic methods for indicating or contraindicating:

Chinese medicine formulas for the treatment of certain arrhythmias can be of more benefit during the prevention stages, especially when the patient is already found to be predisposed to development. Pharmaceutical drugs for these symptoms are already known to cause symptoms such as edema of the ankles/feet, fatigue and weight gain, dyspnea and dizziness, complexion changes and skin rashes, nausea and severe abdominal pain, slowed heart rate, and anaphylaxis. Though Western medicine physicians are aware of the side effects when prescribing, often the drugs are considered to have benefits that outweigh the side effects.

- Bile acid sequestrants interfere with the use of herbal formulas and nutritional supplements. One or the other must be used exclusively. Bile acid sequestrants prevent fat absorption in the intestine and assist in the rise of HDL levels. However, they can also block the absorption of vitamins A, D, E and K, vitamins C and E, and beta carotene.
- Niacinamide is the constituent in niacin which helps in the treatment of low levels of HDL. Certain problems associated with improper use include flushing, increase in histamine, arrhythmias, blood sugar problems, low blood pressure, stomach ulcer, and liver, gall bladder and kidney disease complications.
- Fibrates increase the effect of digoxin and medications for endocrine disorders and should be taken an hour before or at least 4 hours after these types of medication.

(b) Basic methods for prescribing exclusively or integrating:

For patients who elect to use a preventative method, a set lifestyle which minimizes risk factors that lead to qi and blood deficiency, stagnation and stasis is recommended. When Chinese medicine is necessary as part of the prevention program, it is usually to help recover after an exterior invading illness or infection to help nourish and invigorate qi and blood.

- Chinese medicine formulas for the treatment of lipoprotein disorders may benefit patients intolerant of pharmaceutical drugs. It is not recommended that patients blend herbal formulas with drugs owing to certain constituents, including the mineral and vitamin content of the herbs.
- Chinese medicine formulas are as effective as pharmaceutical drugs for lowering blood lipids. In some cases where a patient has been taking drugs to lower lipids and is indicated for Chinese medicine, the patient can be weaned off the drugs and started on Chinese medicine within a few days.
- Shen qi wan, Gentian Formula, Gentian Formula and Fleece flower formula can help to protect viscera with niacin use. Aspirin may reduce the flushing caused by niacin.

Additional Herbal Modifications (see Chapter 19):

With these formulas, herbs from the cardiovascular category, palliative category, antimicrobial category, and blood category may be necessary.

Chapter 12: Sick Sinus Syndrome (SSS)

1. *Pharmaceutical Drugs and Treatments*

(a) Pacemaker: artificially adjusts the atrial heart rhythm.
(b) Radiofrequency ablation: destroys/ablates the atrio-ventricular tissue around the AV node to control the rates of fibrillation in the atria from affecting ventricles.
(c) Antiplatelet medicine: inhibits formation of thrombi, especially to prevent the development of clotting around pacemakers and similar devices, and the development of clotting as a risk factor for stroke.
(d) Beta blocker: blocks the binding receptor on vascular smooth muscle cells that respond to epinephrine, the stress response.
(e) Calcium channel blocker: antihypertensive that decreases blood pressure.
(f) Cardiac glycoside: digoxin may be used for arrhythmias and in conjunction with a pacemaker.

2. *Chinese Medicine Formulas*

(a) Gui zhi tang. Warms to body. Dosage: 8 oz/227 g or 30 tea pills 3 × daily as needed.

(b) Shen fu decoction. Helps the patient to recover after illness, warms yang. Dosage: 8 oz/227 g or 30 tea pills 3 × daily as needed.
(c) Gui pi tang. Nourishes and tonifies blood. Dosage: 8 oz/227 g or 30 tea pills 3 × daily as needed.
(d) Ba zhen yi mu tang. Nourishes and tonifies blood. Dosage: 8 oz/227 g or 30 tea pills 3 × daily as needed.
(e) Zhi gan cao tang. Helps the patient to recover from illness, regulates heart functioning. Dosage: 8 oz/227 g or 30 tea pills 3 × daily as needed.
(f) (f) You gui zhen zhu decoction. Tonifies kidney yin and calms the spirit. Dosage: 8 oz/227 g or 30 tea pills 3 × daily as needed.

Integrating Chinese Herbal Formulas and Western Medicine Drugs During Treatment

(a) Basic methods for indicating or contraindicating:

Patients with sick sinus syndrome are usually elderly patients with chronic atrial fibrillation, which is characterized by alternating patterns of tachycardia and bradycardia. This condition puts the patient at risk for congestive heart failure, stroke and sudden death.

- Generally Western medical health providers refer these patients for a pacemaker, antiplatelet medication and a beta blocker.
- Herbal formulas may be indicated for patients in the prevention and mild symptoms stages, not using a pacemaker.
- Herbal medicine may not be indicated for patients on a pacemaker, and/or is being monitored for sudden death.

(b) Basic methods for prescribing herbal formulas exclusively or integrating:

Chinese herbal medicine may be used during a prevention or recovery program, and can help nourish and invigorate qi and blood.

- Certain formulas – such as shen fu decoction, gui zhi tang and zhi gan cao tang – can be combined to help warm and raise yang, which is often necessary in a course of treatment for recovery of SSS patients after an illness.
- Ba zhen yi mu tang, you gui zhen zhu decoction and liu wei di huang wan nourishes and tonifies blood, tonifies kidney yin and calms the spirit, and may be regulated as a general tonic, which can be modified with a lifestyle adjustment and used over the long term to nourish health.

Additional Herbal Modifications (see Chapter 19):

With these formulas, herbs from the cardiovascular category, antimicrobial category, immune category, blood category and palliative category may be necessary.

Chapter 13: Angina Pectoris

1. *Pharmaceutical Drugs*

(a) Nitroglycerine: acts as an instant vasodilator during a myocardial infarction.
(b) Beta blocker: blocks the binding receptor on heart, kidney, arteries, smooth muscle cells that respond to epinephrine, the stress response.
(c) Calcium channel blocker: antihypertensive that decreases blood pressure.
(d) ACE inhibitor: vasodilator that lowers blood pressure.
(e) Antiplatelet: inhibits formation of thrombi.
(f) Anticoagulant: inhibits formation of thrombi.
(g) Aspirin: antiplatelet drug.

2. *Chinese Medicine Formulas*

(a) Xu ming decoction. Clears heat and activates blood flow. Dosage: 8 oz/227 g or 30 tea pills 3 × daily as needed.
(b) Xue fu zhu yu tang. Moves blood, promotes circulation, stops pain. Dosage: 8 oz/227 g or 30 tea pills 3 × daily as needed.
(c) Tian ma gou teng yin. Calms liver, clears heat, tranquilizes the mind. Dosage: 8 oz/227 g or 30 tea pills 3 × daily as needed.
(d) Gui zhi tang. Relieves the exterior, harmonizes wei qi and ying qi. Dosage: 8 oz/227 g or 30 tea pills 3 × daily as needed.
(e) Pulse-activating powder. Replenishes qi, nourishes and astringes yin. Dosage: 8 oz/227 g or 30 tea pills 3 × daily as needed.
(f) Dan shen dripping pills. Promotes blood circulation, removes stasis, relieves pain. Dosage: 8 oz/227 g or 30 tea pills 3 × daily as needed.

Integrating Chinese Herbal Formulas and Western Medicine Drugs During Treatment

(a) Basic methods for indicating or contraindicating:
- The use of Chinese herbal medicines for patients with angina pectoris requires timing. Prevention and control of symptoms during the uncertain diagnoses and stable angina stage is necessary to avoid future complications.
- Coronary artery disease is a key concern with angina pectoris. Health care providers considering integrative therapies in treatment should evaluate the type of angina the patient is experiencing, and whether herbal medicine might be indicated in the approach to treatment. For example, unstable angina attacks may not always stabilize with medication, but still respond to sublingual nitroglycerine. By the unstable stage, Western medicine health providers have already prescribed and advised the patient to keep nitroglycerine nearby in cases of acute myocardial infarction.

- Variant angina, which often occurs at night and can be fatal, may be beyond the limitations of herbal medicine in most patients.
- Microvascular angina occurs primarily in women; detection is difficult using PCI and angiograms, and requires close observation of the patient's lifestyle habits, regular lipid profile, hormonal testing and narrative information to determine treatment success.
- Patients not responding to Chinese herbal medicine alone may eventually need to transition to Western medicines, by integrating aspirin once daily. If, and as, symptoms become more unstable, patients may need to begin beta blockers, calcium channel blockers, ACE inhibitors, anti platelet medicine and, perhaps, an anticoagulant.

(b) Basic methods for prescribing exclusively or integrating:

- Generally, during the prevention stages, patients may respond well to Dan shen dripping pills. During an episode of palpitations, the patient can place 10 micro-pellets under the tongue, and within minutes the relief is felt, with a cool and comfortable sensation in the chest. Tian ma gou teng yin is a clinical standard to control hyperactivity, and can be modified easily. Some patients with insomnia and night-time discomfort may require dosages in the evening to assist with sleep.
- Pulse-activating powder can also be prescribed as a final prevention effort where the patient may need to recover from illness caused by exterior pathogens and advancement of angina symptoms with profuse perspiration. This medicine helps replenish qi and astringe yin.
- Gui zhi tang, xue fu zhu yu tang and xu ming decoction, relieves the exterior, harmonizes wei qi and ying qi, moves blood, promotes circulation and stops pain, and should be modified as part of a seasonal regimen anytime during the year again various microbial infections which cause colds and influenzae.

Additional Herbal Modifications (see Chapter 19):

With these formulas, herbs from the cardiovascular category, palliative category, antimicrobial category, immune system, and blood category may be necessary.

Chapter 14: Rheumatic Fever

1. Pharmaceutical drugs

(a) Aspirin: anti-inflammatory, analgesic, antipyretic.
(b) Corticosteroids: anti-inflammatory.
(c) Penicillin: antibacterial.
(d) Erythromycin: antibacterial.
(e) Azithromycin: antibacterial.

2. *Chinese Medicine Formulas*

(a) Qing wen bai du yin. Treats severe toxic heat. Dosage: 8 oz/227 g or 30 tea pills 3 × daily as needed.
(b) Long dan xie gan tang. Treats damp heat which affects the lower jiao. Dosage: 8 oz/227 g or 30 tea pills 3 × daily as needed.
(c) Shen tong zh yu tang. Unblocks meridians and moves qi and blood. Dosage: 8 oz/227 g or 30 tea pills 3 × daily as needed.

Integrating Chinese Herbal Formulas and Western Medicine Drugs During Treatment

(a) Basic methods for indicating or contraindicating:

- Chronic rheumatic fever is often characterized by a temperature of more than 101°F (38.2°C), malaise, joint pain, subcutaneous nodules around the joints, and, in some patients, a body rash called erythema marginatum. Symptoms are often managed by offering the patient antibiotics, including sulfa drugs. Corticosteroids and anti-inflammatory medications are prescribed for the painful joints.
- In some patients, the long-term use of antibiotics will cause secondary yeast infections, a weak immune system, and the risk of noncompliance. Some patients may experience severe side effects, especially on the corticosteroid medication.
- Some patients may benefit from integrating Western medicine antibiotics with Chinese herbal antibiotics by substituting sulfa drugs in some patients with an allergy to or low tolerance for side effects.

(b) Basic methods for prescribing herbal formulas exclusively or integrating:

- Health practitioners treating patients with strep throat or toxic heat pathogens may find it necessary to provide liquid decoctions over a relatively long period of time (5–7 days) while at the same time cooperating with Western medicine pharmaceutical drugs. The flavor of antibiotic Chinese medicine decoctions is bitter and strong, but the decoctions are effective. A suggestion may be to monitor the patient's condition weekly, introducing yang-tonifying herbs until blood work shows that the body has eliminated the microbe.
- Some patients may benefit from integrating the Western medicine antibiotics with Chinese herbal antibiotics. Qing wen bai du yin may be modified and substituted for sulfa drugs in some patients with an allergy to, or low tolerance of, sulfa drugs. Also, the herbal formula can be prescribed to help lower the high fever and the toxic heat pathogens. In some cases it may be used alone. Prescribing herbal formulas used in conjunction with antibiotic drugs, will require the condition to be monitored with laboratory tests. In some patients,

aspirin may be necessary to continue over the long term to thin the blood, especially during the recovery period after a severe illness. The aspirin, along with modifying a recovery formula, such as shen fu decoction, can help patients after an illness.

- Formulas such as long dan xie gan tang can be modified to provide relief for lower jiao complications such as gastrointestinal and urinary problems associated with the toxic heat. Shen tong zhu yu tang can be modified to move qi and blood and dredge the meridians to address the joint and limb pain.

Additional Herbal Modifications (see Chapter 19):

With these formulas, herbs from the cardiovascular category, antimicrobial category, immune system, anti-pyretic, and blood category may be necessary.

Chapter 15: Endocarditis

1. Pharmaceutical Drugs

(a) Intravenous vancomycin: *Enterococcus* infection.
(b) Intravenous ceftrioxene: resistant bacterial infections.
(c) Penicillin: Enterococci, Gemella sp., Granulicatella sp. and *Abiotrophia defectiva.*
(d) Nafcillin: *Staphylococcus aureus.*
(e) Aminoglycosides: antibiotics which treat aerobic Gram-negative bacterial infections.
(f) Intravenous benzyl penicillin: *Streptococcus bovis.*

2. Chinese Medicine Formulas

(a) Sheng yang yi wei tang. Resolves exterior pathogenic factors, drains dampness, raises yang. Dosage: 8 oz/227 g or 30 tea pills 3 × daily as needed.
(b) Ren shen tang. Replenishes qi, nourishes yin, raises yang, treats palpitations and vertigo. Dosage: 8 oz/227 g or 30 tea pills 3 × daily as needed.
(c) Qing wen bai du yin. Clears toxic heat from blood, nourishes yin. Dosage: 8 oz/227 g or 30 tea pills 3 × daily as needed.
(d) Pu ji xiao du yin. Clears toxic heat and wind damp-heat, reduces swelling. Dosage: 8 oz/227 g or 30 tea pills 3 × daily as needed.
(e) Long dan xie gan tang. Clears excess liver and gall bladder damp heat. Dosage: 8 oz/227 g or 30 tea pills 3 × daily as needed.
(f) Shen tong zhu yu tang. Activates blood circulation and qi flow, dredges meridians and collaterals and relieves pain. Dosage: 8 oz/227 g or 30 tea pills 3 × daily as needed.

Integrating Chinese Herbal Formulas and Western Medicine Drugs During Treatment

(a) Basic methods for indicating or contraindicating:

- Because endocarditis is a potentially life-threatening cardiovascular condition, Western medicine health providers may restrict treatment to antibiotic therapy.
- Integrative health care providers and patients who elect to use herbal therapy may follow the suggestions from Chapters 14 and 15. The toxic heat pathogens should be treated as early as possible, especially during rheumatic fever or scarlet fever. In some cases it may be necessary to provide liquid decoctions from both chapters, either isolated or combined and modified over a longer period of time (12–30 days).
- In some patients, the long-term use of antibiotics will cause secondary yeast infections, a weak immune system, and the risk of non-compliance. Some patients may experience severe side effects, especially on the corticosteroid medication.
- Some patients may benefit from integrating the Western medicine antibiotics with Chinese herbal antibiotics by substituted sulfa drugs in some patients with an allergy or low tolerance for side effects.

(b) Basic methods for prescribing exclusively or integrating:

- Health practitioners treating patients diagnosed with toxic heat pathogens may find it necessary to provide liquid decoctions over a relatively long period of time (12–30 days) while at the same time alternating by taking 2 hours before any Western medicine pharmaceutical drugs. The formulas qing wen bai du yin, pu ji xiao du yin, long dan xie gan tang and shen tong zh yu tang can be combined and modified into a custom formula during treatment.
- The flavor of these and most antibiotic Chinese medicine formula decoctions is bitter and strong, but the decoctions are effective. A suggestion may be to monitor the patient's condition weekly to help monitor the stomach from the effects of the cold herbs until blood work shows that the body has mostly eliminated the microbes.
- Finally, begin introducing a modified decoction of sheng yang yin wei tang for yang-tonifying and blood-moving and yin-nourishing herbs, while adding a few extra antimicrobial herbs to the decoction for an additional 6 months to 1 year.
- In addition, some patients may need a modified decoction or extract containing only, or primarily, ren shen, such as shen tang, ren shen tang or ba zhen yi mu wan to help recover while addressing any signs of blood deficiency, while monitoring the progress on CBC blood results.

- As early as possible during treatment, encourage the patient to organize diet and lifestyle modifications which can further personalize the treatment plan.

Additional Herbal Modifications (see Chapter 19):

With these formulas, herbs from the cardiovascular category, antimicrobial category, immune system, anti-pyretic and blood categories may be necessary.

Chapter 17: Thromboangiitis Obliterans (TAO)

1. Pharmaceutical Drugs

(a) Hyperbaric oxygen therapy: for vasodilation.
(b) Prostaglandins: for vasodilation.
(c) Streptokinase: for antithrombosis.
(d) Corticosteroids: anti-inflammatory.

2. Chinese Medicine Formulas

(a) Tao hong si wu tang. Breaks blood stasis, moves blood. Dosage: 8 oz/227 g or 30 tea pills 3 × daily as needed.
(b) Si miao yong an tang. Treats gangrene. Dosage: 8 oz/227 g or 30 tea pills 3 × daily as needed.
(c) Wu wei xiao du yin. Clears toxic heat and toxins. Dosage: 8 oz/227 g or 30 tea pills 3 × daily as needed.
(d) Xian fang huo ming yin. Clears toxic heat and toxins. Dosage: 8 oz/227 g or 30 tea pills 3 × daily as needed.
(e) Tou nong san. Removes toxins, replenishes qi and blood. Dosage: 8 oz/227 g or 30 tea pills 3 × daily as needed.
(f) Yang he tang. Treats gangrene. Dosage: 8 oz/227 g or 30 tea pills 3 × daily as needed.

Integrating Chinese Herbal Formulas and Western Medicine Drugs During Treatment

(a) Basic methods for indicating or contraindicating:
  - TAO symptoms are characterized by chronic and frequently recurring stages of inflammation and thrombosis of the legs and feet. It is common in young males and is thought to occur as a result of tobacco use, so helping the patient with side treatments for tobacco cessation is crucial.
  - As in Western medicine, treatments using Chinese medicine should be seen as a palliative care option which mostly addresses symptoms.

Chinese medicine practitioners and Western medicine practitioners should collaborate with the patient to document medicines and methods which have helped to make the most progress and prolong remission periods.

- Herbal medicine and pharmaceutical drugs may be able to work together because they will address different issues in the patient's condition. Western medicine drugs may address vascular complications, while Chinese medicine can address the purulent lesions with medicines at each stage and can be prepared for ingestion and external application.

(b) Basic methods for prescribing exclusively or integrating:

- Personalizing treatments for patients with TAO requires patience and acceptance that the efforts are palliative, and the lifestyle may need to include home remedy projects to gain optimal relief. Blood work to monitor inflammation and microbial infection is necessary to help adjust treatments.
- Prepare *two* batches of liquid decoctions – for internal use and external use.
- For internal use, modify the formulas according to differentiation, and include other herbs which balance hormones as well. The medicine can be made into capsules, teas, keli granules and extracts.
- For external use, the decoction can be processed with a natural wax and good plant-based oils and made into ointments and creams. In addition, it can be added to a favorite liquid soap or processed with a saponification substance and good plant-based oils and made into a bar soap for external wash.
- The disease also develops during three periods. During the ischemic period, the veins of the lower leg are inflamed, and the skin has ulceration and gangrene, with intermittent claudication. Tao hong si wu tang, si miao yong an tang and wu wei xiao du yin can be taken internally. Si miao yong an tang, tou nong san and yang he tang can be made into an external preparation.
- During the nutritional disorder period there is persistent pain and muscle atrophy. Tao hong si wu tang si miao yong an tang Wu wei xiao du yin Yang he tang can be taken internally. Si miao yong an tang, tou nong san and yang he tang can be made into an external preparation.
- During the necrosis period, the pain becomes worse as infection leads to the development of ulcers and gangrene. Tao hong si wu tang, xian fang huo ming yin and Tou nong san can be taken internally. Si miao yong an tang, tou nong san and yang he tang can be made into an external preparation.

Additional Herbal Modifications (see Chapter 19):

With these formulas, herbs from the cardiovascular category, antimicrobial category, immune system, anti-pyretic, endocrine and blood category may be necessary.

Chapter 18: Takayasu Arteritis (TA)

1. Pharmaceutical Drugs

(a) Corticosteroids: help control symptoms during the inflammatory process.
(b) Cytotoxic agents: immunosuppressant medication combined with steroids for advanced stages when relapses are more frequent.
(c) TNF inhibitors: combine with corticosteroids and cytotoxic agents during remission stages to decrease immune responses.
(d) Antiplatelets: inhibit platelet aggregation causing thrombosis, and renal failure.
(e) Anticoagulants: reduce or block further thrombosis.
(f) Calcium channel blocker: treats hypertension.

2. Chinese Medicine Formulas

(a) Si miao yong an tang. Clears heat, detoxifies and activates blood. Dosage: 8 oz/227 g or 30 tea pills 3 × daily as needed.
(b) Wen yang tong mai tang. Activates blood, relieves pain, invigorates deficiency. Dosage: 8 oz/227 g or 30 tea pills 3 × daily as needed.
(c) Tong mai huo xue tang. Activates blood and removes stasis, regulates flow of qi, replenish blood. Dosage: 8 oz/227 g or 30 tea pills 3 × daily as needed.
(d) Bu yang huan wu tang. Invigorates qi, resolves blood stasis. Dosage: 8 oz/227 g or 30 tea pills 3 × daily as needed.
(e) Bu pian yu feng tang. Replenishes qi, soothes liver and stops wind. Dosage: 8 oz/227 g or 30 tea pills 3 × daily as needed.

Integrating Chinese Herbal Formulas and Western Medicine Drugs During Treatment

(a) Basic methods for indicating or contraindicating:
- TA symptoms through certain phases can become complicated and vague to discern in some patients. The method of integrating medicine will require timing, close observation and deep cooperation between Chinese medicine practitioners, Western medicine practitioners and the patient.

- The Western medicine view of TA recognizes three phases – early, pre-pulseless and late – which is characterized by vascular complications (especially the aorta), a diminishing pulse and development of heart failure.
- When integrating Chinese medicine and Western medicine, it is necessary to realize the side effects and complications caused by the pharmaceutical medications. In some cases the strategy for integrative medicine treatment may include using Chinese medicine during the early and pre-pulseless phases. Western medicine can be used during the active and late phases. Chinese medicine can be used again in the stable phase during remission.
- During active or stable phases, it is suggested that herbal formulas be contraindicated if anticoagulant medications are used, because they provide equal pharmacodynamic action. If the formulas are used, they must be profoundly modified for compatibility with the course of treatment.

(b) Basic methods for prescribing exclusively or integrating:

- Treatment using Chinese medicine can work with Western medicine to relieve physical symptoms such as pain, weakness and emotional disturbances. Because it has been documented that interleukin-12 is involved in inflammatory disease, Chinese medicine formulations can be provided to address inflammation during remission or active phases.
- During the active phase, si miao yong an tang, da qing dao tang, wen yang tong mai tang and tong mai huo xue tang can be used to assist in sustaining the patient by invigorating and moving the blood. The formulas can be modified and combined into a single decoction which acts as a nutritional supplement and is alternated with the TNF inhibitors and cytotoxic agents.
- During the stable phase, tong mai huo xue tang, bu yang huan wu tang, and bu pian yu feng tang assist in replenishing deficiency in the patient recovering from the active phase. The formulas can be modified and combined into a single decoction which acts as a nutritional supplement and is alternated with the TNF inhibitors, cytotoxic agents, and other medicines which may be suggested during remission. Also, the Chinese medicines can be used alone without the aid of pharmaceutical drugs. However, very close observation is necessary.

Additional Herbal Modifications (see Chapter 19):

With these formulas, herbs from the cardiovascular category, antimicrobial category, immune system, anti-pyretic, endocrine and blood category may be necessary.

Index

Note: Page numbers followed by "*f*" and "*t*" refer to figures and tables, respectively.

D

E

F

G

H

I

T

U

V

W

X

Y

Z

FSC
www.fsc.org
MIX
Papier | Fördert gute Waldnutzung
FSC® C083411

Kolibri 360 GmbH
Ferdinand-Jühlke-Straße 7
99095 Erfurt, Deutschland
produktsicherheit@kolibri360.de